THE SMARTPHONE WALLET

UNDERSTANDING THE DISRUPTION AHEAD

A TLG Book
Hamilton Building
43 West Front Street 14-A
Red Bank, New Jersey 07701

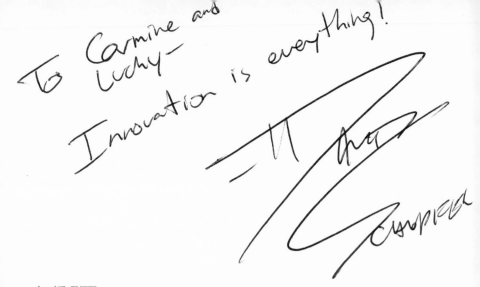

To Carmine and
Lucky—
Innovation is everything!

THE

SMART**PHONE**

WALLET

Understanding the Disruption Ahead

David W. Schropfer

A TLG BOOK
Printed in the USA
Charleston, South Carolina

TLG

Hamilton Building
43 West Front Street 14-A
Red Bank, New Jersey 07701

20   19   18   17   16   15   14
40   39   38   37   36   35   34   33   Pbk.

Schropfer, Jr., David W
*   The SmartPhone Wallet:*
Understanding the Disruption Ahead / David
W. Schropfer, Jr.

ISBN-13: 978-1456429973
ISBN-10: 1456429973

TO MY WIFE
Inspiration Incarnate

Acknowledgements

This work began in 1984 when, as a high school junior, I sold my first cellular phone while working after school for RadioShack. The device weighed over 12 pounds, was over 1 foot long and 1 foot tall, and had a built-in handle so it could be carried like a suitcase.

In the decades that followed, I watched the cellular phone evolve into a smarter, faster, and significantly smaller device. I am truly grateful to the people who, over those years, helped me better understand this technology, its inner workings, and economic impact. These people include: John Cate, Ted Urban Jr., Jody Cipot, Louis Provost, David Tyler, Moshe Kagenoff, Meg Collin and many others.

More recently, I have had the pleasure of working and debating with other seasoned professionals and thought leaders in all three industries (telecommunications, payments, and loyalty) that are coming together thanks to the revolutionary power of the SmartPhone. They include: Lynne Gallagher, Mike Biamonte, Bob Egan, Mike Catalano, Philip Andreae, and others. I look forward to more discussions with you all in the future.

I am sincerely grateful for the support of my partners at the Luciano Group: Tom Luciano, Tom Bruhl, Simon Krieger, Charlie Meyers, Bryan Pearson, and Jeff Needel.

I probably would not have entered the world of technology without the consistent support of my parents, David Sr. and Gloria. Only as dedicated a pair as they could have persevered through my early years of electronic curiosity when I frequently caused mild damage to our home with my experimentation – including almost burning down the house at age eight in a self-taught discovery of the subtle differences between direct and alternating currents.

My friend and colleague, John Greenberg, has played several key roles in the development of this book. A lifetime sounding board and confidante, John's views on the economic problems of the developing nations of the world seemed solvable with technology, and became a part of my work.

Finally, this book would not exist without the unwavering support of my wife, Colleen. There is no possible way to detail the depth and breadth of her support, so I will just say that she was my inspiration, my motivation and my editor. My deepest thanks to her, especially.

Contents

CHAPTER 1: THE VISION   17

CHAPTER 2: THE BREAKTHROUGH   29

CHAPTER 3: THE FORMAT WAR   47

CHAPTER 4: EVOLUTION OF PLASTIC   61

CHAPTER 5: AROUND THE WORLD, TODAY   97

CHAPTER 6: WHAT'S NEXT?   111

CHAPTER 7: CONCLUSION   129

CHAPTER 1:

THE VISION

T he year is 2015. You walk into Wal-Mart. The first thing you see is a large marketing poster with a big bulls-eye painted on it. In the center, it says: "Tap here for today's deals and discounts." You tap your SmartPhone to the bulls-eye almost without thinking. Your phone is now loaded with details regarding what's on sale, what is out of stock and new coupons for your use on that day.

You finish your shopping and head to checkout. There, you find another bulls-eye on the payment terminal, and again you tap your SmartPhone there. What happens next is truly amazing:

Your SmartPhone lists all of your individual payment types for you. That's clever enough, but how do you think this list is organized? Alphabetically? Most frequently used? No.

Instead your SmartPhone shows you a list of each payment type at your disposal based on the **amount of the discount** that you will get for using each type! For example:

Maybe you tap your finger on the name of each account to remind yourself about the interest rates and terms, then choose Chase MasterCard to get your 3% discount. Next, your SmartPhone wants to make sure that you are you. So, it asks for your four-digit PIN number. Or, it may scan your thumbprint. Retina-scan? Maybe. But the PIN is much easier.

Regardless of how you enter your security information, you will probably do it on your SmartPhone, not on the keypad at the checkout counter.

This is a glimpse into one possible future. Of course, there are other visions of how these new systems will work, and the reality of these systems when they mature could be, in fact, even more convenient than what is described here.

One thing is for sure: the future of mobile commerce, powered by the mobile wallet, is a future of convenience, security, and an empowered consumer, and it is also a future that you can help shape and define. In fact, you, the consumer, will have to build it. This book will help you understand what that future looks like, how you can influence it and why every consumer can have an impact on the final results.

Its also important to note that while the benefit of being able to handle money more effectively may lead to greater convenience in the developed world; it is also true that in developing countries it has the potential to turn around stagnant economies. For example, although mobile payments are an accepted and growing form of transacting money in Kenya, Nigeria, South Africa, and other countries in Africa, it was recently outlawed in Somalia. Why? Well, it depends on who you ask -- and of course there are many opinions -- but the opinion of this author is that paying with a SmartPhone creates a paper trail, just like paying with a credit card. Cash, of course, does not leave a paper trail. So, if there is corruption among the current leaders of the tragic nation of Somalia, then outlawing mobile payments would ensure that that the system of corruption is maintained. But, that is just a guess.

What is not a guess is that history is littered with great ideas that have died in the marketplace. Cause of death number one is usually because another product with better marketing, and better publicity, and inexplicably more users, simply buries a superior product. As a result, we all lose the benefit of the superior product, and we have to settle for the mediocrity that we, the consumer, chose.

Right now, all over the world, a format war is raging over the evolution of how we all handle our money. Whether we are managing our own accounts, paying a retailer for their goods, paying a hotdog vendor for his "food", paying a friend for your half of the dinner tab -- there are tens of thousands of people already designing and building a more efficient system based on our most modern and most utilized communications tool: the SmartPhone.

The choices we all will be asked to make in the very near future will affect which version of the mobile wallet becomes the standard that we all will have to use.

Bottom line: if you want these changes to occur, pay close attention to the choices you make regarding any mobile wallet or mobile commerce product that you use over the next few years. And, keep an open mind. If you choose a given product, and later find another that better suits your needs, be willing to try that new product - especially if you would like to have that new product around for the next couple of decades. We are right at the beginning of a new era and we all have the opportunity to affect the end results, but we must keep an open mind and stay informed. There will be many options to choose from and as consumers we have the power to get the best, most secure, easy-to-use solution possible, but only if we are paying attention.

If you like to try new products and services as soon as they are available, then you can use the technology today – right now. This book will tell you how to do it -- Safely. For everyone else who just wants to know what is happening, when the technology will be in the mainstream, and what benefits it will offer, this book will also explain all of that in a language that industry outsiders can easily understand.

Practical Terminology

This is not a technical manual; the reader does not need an engineering degree to understand this book. You only need to be alive, and a consumer of goods to understand the concepts and the advice in this discussion.

However, if you have questions about any of the terms in this book, this short glossary should help.

SmartPhone

In this book, the term "SmartPhone" means a mobile phone with at least one of these characteristics:

(1) Any mobile phone with an Internet browser.

(2) Any mobile (or "cellular") handset that runs a complete operating system capable of running multiple applications (also called 'software programs.') are commonly referred to as "SmartPhones."

(3) Any phone on a GSM network (2.5G, 3G, or any of the variations of 4G), as long as the phone is capable of running

simple applications (or "applets"), including applications written for platforms such as Java ME. These types of mobile phone handsets are also called GSM "feature phones." If you use a mobile phone on a GSM network, then (with only extremely rare exceptions) your phone will be capable of utilizing at least one of the forms of a SmartPhone Wallet. Why? [This requires some technical terms to describe.] In this book, you will find the description of a new capability called a 'Trusted Device' that is enabled by a new player in the value chain of mobile phones called a Trusted Service Manager (TSM), and utilizes the core component of a GSM Phone: the removable Universal Integrated Circuit Card (UICC) which is able to contain several different applications. The primary application of the UICC is the 'subscriber identification module' (or "SIM") that contains the secret, randomly generated codes that matched the phone's certification that the GSM operator provided when the phone was first authorized on mobile operator's network. As a result, the UICC is often referred to as a "SIM Card". But, the UICC can do much more than just the SIM functions; it can be used to certify the device on other types of secure networks also. This capability could enable any GSM phone to become a secure and 'trusted' device on many different types of networks. The UICC Card can also include a small antenna to enable a feature phone to be capable of near-field communications (NFC). These are all advanced functions that, if utilized, would more than earn the title of 'SmartPhone" to any ordinary feature phone.

SmartPhone Wallet

When you purchased the physical wallet that is in your pocket or purse, right now, the wallet was empty at first. Then, you chose to put things in your wallet, such as credit cards, loyalty cards, and coupons. The SmartPhone Wallet also is technically empty when you buy it, and then, over time, you decide to add things to it, such as bank accounts, credit accounts, loyalty accounts, etc.

For the purpose of this book, that is the general definition of the SmartPhone Wallet, namely "an empty electronic vessel" which can be filled with various types of accounts, rewards, discounts and other services which are provided by a variety of different companies.

As of the end of 2010 when this book was written, the leading three forms of a SmartPhone Wallet included:

1) **A SmartPhone Handset is the "Empty Vessel:"** In this scenario the customer selects individual applications from individual service providers (such as credit card issuers, loyalty programs, retail stores, etc). This will give the customer a high degree of customization, but be generally more difficult to manage. Also, each application would have no ability to work in conjunction with any other application, unless both were specifically written to do that.

Open Revolution's
Mobile Web Wallet

Source: Open Revolution, Used by Permission

2) **A Mobile Web Browser is the "Empty Vessel:"**

– The browser could load a secure web site (or web sites) that would allow you to perform various payment, rewards, and other functions. This method is dependant on the ability to connect to the Internet from your mobile phone, which can be less secure and sometimes too slow to use to make a purchase. The main advantage, however, is virtually unlimited functionality from the web site because the computing power and computing speeds are not limited by the computing power of the SmartPhone or its Mobile Web Browser.

3) **An Application Specifically Written to Manage Accounts Issued by, or Authorized by, Third Parties is the "Empty Vessel:"** In this case, the application is specifically written to collect the functionality of other accounts and execute transactions from these other accounts from a single application. While this format presents the best possible convenience for the end user, and a high potential degree of integration (such as the example of loyalty and payment integration described in the example at the beginning of this introduction), the downside theoretically is a third party managing your individual sensitive account information for you. Unless the third party provided a high degree of trust, then this method may be less attractive to some consumers.

Privacy and security are the lynchpins of all of these types of cards. If either the privacy or the security are not precisely executed, the potential exists for a compromised system.

This book explains how the industry is going to address these concerns, but it is up to the individual consumer to decide if they are comfortable with the promises.

V/MC/AMEX

In this book, V/MC/AMEX stands for Visa, MasterCard, and American Express, and their competitors like Discover and JCB. This book would have been dozens of pages longer if each reference to all of these companies relied on repetitively naming each company.

The Payment System

All major credit and debit cards using the brand name of V/MC/AMEX involve a system (or a scheme) of sending money from a consumer's bank (or credit) account to a retailer's bank account. There are a wide variety of differences in how these systems work in various parts of the world, how security is provided, and how the account information is communicated to an electronic cash register (magnetic stripe, contactless, 'chip and PIN', EMV, NFC etc). In general, these are not issues to be addressed in this book because to the consumer, these processes are invisible after their card is swiped or waved or inserted. Therefore, variances in the system are not a critical aspect, and the industrial variances are generally not debated in this book.

Near Field Communications (NFC)

Near field communications (NFC) is just another two–way radio signal that allows one device to communicate with another device. Your SmartPhone already emits more than one different kind of two–way signal, including (in most cases) Bluetooth and Wi-Fi. So, it may seem unnecessary to have another two–way signal coming from your SmartPhone.

However, the significant advantage of NFC is the range.

Both Bluetooth and Wi-Fi are signals that your enabled SmartPhone can connect with from around 10 to 50 feet away, and even more in some cases. When conducting a retail transaction with your phone, you probably don't want your communication signal to carry that far, lest you initiate a transaction without knowing about it by accidentally pressing a button or two on your SmartPhone when you are in the proximity of an electronic cash register.

If you use NFC as the communication channel between your SmartPhone and the electronic cash register, the range is not a problem because the NFC range is approximately 5 centimeters. Your SmartPhone virtually needs to be in contact with the device it is communicating with, which is exactly the point; no unintentional communication.

NFC is most likely to become the standard of choice in securing the communication stream between a SmartPhone and a cash register. Although NFC takes many forms, and exists from many different companies today, they generally take the form of a sticker, and add-on device, or embedded in the form of an electronic component into the SmartPhone itself by the manufacturer.

CHAPTER 2:

THE BREAKTHROUGH

hen was the last time you saw an error on your

mobile phone bill like an expensive call that you didn't make because someone stole your phone number and placed a call from another phone? Not too long ago, this was a common problem, but it is not today. Why? Because mobile phone companies (also known as mobile network operators) figured out the security necessary to prevent this fraud. So now, only your phone can make calls that will show up on your phone bill.

Mobile phones started out essentially like a two-way radio. They were designed so that you could talk to someone, and someone could talk to you. The phone had an encrypted serial number burned into a memory chip, and it transmitted this number to the mobile network to identify itself. But, not long after I sold my first cellular phone (the 12 lb. one I mentioned earlier) at age 17 while working for RadioShack, phone security very quickly became an issue. Some hacker out there figured out how to essentially "record" the serial number that an authorized phone transmitted to the mobile network, then program another phone to transmit the same encrypted number.

So, the phone that stole the serial number could make all the calls it liked, but the owner of the original phone got the bill.

And it was a big bill. At the time, every minute you spoke on that massive cellular phone cost a lot of money; much more than it costs today. So, if somebody somewhere was financially motivated to make the network confused about whom to charge for all that talking they could make all of those charges get billed to someone else's cellular phone. Eventually, even a moderately skilled computer hacker could achieve this goal. Phone companies realized quickly that a new security format was needed.

Fast-forward about 20 years, and the problem is solved. Here's how they did it:

Without getting into the technical muck, most SmartPhones are able to authenticate the subscriber on the mobile network using a "secret" that was shared between the mobile network and the SmartPhone at the time the SmartPhone was first activated on the network. The "secret" is stored on the UICC Card (see the definition of the SmartPhone Wallet in the previous chapter) and then used by the mobile phone to answer a question asked by the network. If the mobile phone provided the correct answer back to the network using the secret, then that phone is authentic. If not, it is blocked.

This means your mobile phone is repeatedly authorized to operate on your mobile network, as if it were a component within the network itself.

So, security issues on cell phones were solved by making the mobile phone a "trusted component" of a network. This means that your mobile phone company recognizes your phone as an extension of its network because the security method is so strong.

Here is the big breakthrough idea.

If your SmartPhone can be a secure part of your mobile phone network, why can't it be a secure part of other networks, too?

This goes far beyond the simple applications on your SmartPhone; it is a new technology that has the potential to change everything you know about handling money, using coupons, accumulating loyalty points, even the form of your personal identification.

Your bank has a network; if your SmartPhone was as secure as an ATM machine on your bank's network, what could you do with it? Could you distribute cash directly from your checking account to a retailer's bank account without using anything like a Visa card or MasterCard?

Yes.

Your credit card company (which may or may not be the same as your bank) has another network: if the SmartPhone in your hand could connect to your credit card company with as much security as the computer terminal of their customer service representatives, would you even need an actual plastic card, or would your phone be a more secure way to ensure that no one can use your account information except for you?

How about your department of motor vehicle's network?

Your SmartPhone can be added to almost any network you like, with the ability to perform almost any function possible on that account, as long as you give your SmartPhone permission. This technology exists right now, and right now, companies are falling over themselves to make sure that they are positioned to play in this new market.

There are no shortage of books in the marketplace that can predict the long-term

> **If your SmartPhone can be a secure part of your mobile phone network, why can't it be a secure part of other networks, too?**

future for some of these new technology applications and benefits. In this book, you will primarily read about retail payments; an action-- that as of the writing of this book at the end of 2010 -- are accomplished primarily with credit cards, debit cards, and cash.

Your Wallet Is Already Electronic

The concept of replacing your physical wallet with your SmartPhone may not be intuitive, unless you consider that everything in your physical wallet today is part of an electronic network.

Let's try an experiment to prove that point:

Take out your wallet. Seriously. If you do, I guarantee that you will completely understand the term "SmartPhone Wallet" as well as any industry pro in the next two minutes.

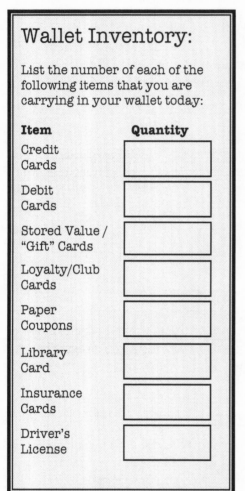

Wallet Inventory:

List the number of each of the following items that you are carrying in your wallet today:

Item	Quantity
Credit Cards	
Debit Cards	
Stored Value / "Gift" Cards	
Loyalty/Club Cards	
Paper Coupons	
Library Card	
Insurance Cards	
Driver's License	

Now, spread everything out on a table, and group the like items together. Write the number in the Wallet Inventory worksheet provided here. The results may surprise you.

The first wallet inventory I completed was on the wallet of a person who shall remain nameless (but gave me permission to do it). This person had two credit cards, three debit cards, eight loyalty "Club" cards, two Gift cards, eight paper coupons, a library card, three insurance cards, a driver's license and some cash.

What do ALL of these items have in common?

Except for the cash, every item in this wallet -- and in your wallet -- is managed by a computer network.

Although this was not always so, clearly today a computer network is involved in the management of every item in your Wallet Inventory.

Credit cards and debit cards are the most obvious. Extremely large, extremely expensive network operating systems govern every request you make for a new transaction at a retailer, or every time you make a payment, or every time your monthly bill is generated. All of your credit card activities are managed by several computer networks working together.

Gift cards -- also known as "stored value cards" -- are managed by computer networks also. With similar levels of security to the credit card and debit card networks, stored value cards are managed by networks that, in most cases, are responsible for determining the current balance of your card at any given time. Some stored value cards actually record the balance of the card on the electronic elements within the card, however, even in those cases, a computer network is responsible for validating the card each time you use it.

Loyalty cards, also known as "points cards," "rewards cards," or "club cards" are managed by computer networks that track the amount of money you spend on every qualifying transaction to determine how many points (or discounts or other rewards) that you have earned based on each individual program. Those same networks are also responsible for keeping track of each time you use (or "burn") your rewards. These programs generally are not used to purchase goods and services. Instead, these programs generally require you to accumulate enough value on the card so that you may then trade that value for something specific, for example buying a DVD player for 10,000 points on your local grocery store "club card." Regardless of the specific form or function of your rewards program, the issue here is that all activity on the rewards account which is represented by the card in your wallet is managed, tracked, and secured by a computer network.

Paper coupons are not as obvious but they are also managed by a computer network. Why? Because retailers, and other companies that issue coupons, are highly interested in the offers that you respond to, and do not respond to. The two extremes are a coupon that you cut out of a newspaper, versus a coupon that was mailed directly to you at your home. In the case of the newspaper, there is obviously no way for the retailer to know very much about you as an individual when you present that coupon at the cash register for your discount. However, all of the other information that can be tracked -- is tracked. For example, the name of the newspaper that printed the coupon, the date the coupon was printed, the date the coupon was used, where the coupon was used, the purchase amount of the transaction when the coupon was used. And in some cases, for those retailers with more advanced networks, they can track that you as an individual used the coupon based on the credit card or debit card that you used to pay for the transaction. Note: when they track "you" in this case, they are really tracking the card you used to make a purchase, and the "profile" does not, and (based on industry standards) cannot, include personal information other than possibly your name unless you authorize the retailer to do so.

> **Everything in your wallet, except cash, is somehow managed by a computer.**

On the other hand, when a coupon is sent to you at your home, a computer network has stored the barcode that is printed on your coupon so that the computer network knows when, where, and how much you spent when you redeem that coupon. In this case, obviously, the retailer does have your name and address and is compiling a profile of you for your – and their -- benefit.

The remaining items on your SmartPhone Wallet inventory are also tracked, managed, and secured by the computer networks that issue them, including the library cards, insurance cards, and your driver's license.

Privacy

This is a good time to digress briefly into the topic of privacy. So, while the purpose of this book is to inform you about the "disruption" that will happen over the next few years related to your SmartPhone, and how every item you currently carry in your wallet will migrate to a SmartPhone environment; privacy continues to be a major issue for all consumers.

The previous section described how a computer network is involved with virtually every item in your physical wallet, and it can be unsettling to know how many different computer networks are aware of who you are and what you are doing at any moment in time and this does not even count the other networks that track your activity every time you go on to the Internet, or log into a computer. But that is a topic for another book.

What you need to know about privacy is that using the "SmartPhone Wallet" may in fact give you a chance to start over with a clean slate.

Here is why: they need your permission.

Every database that tracks you as a consumer today received your permission to do that in one form or another. So, keep that in mind as you start using applications that will enable your SmartPhone to replace different elements in your wallet.

> Moving from your existing wallet to the SmartPhone Wallet may give you an opportunity to regain your privacy.

If you choose a loyalty program that operates from your SmartPhone, or from your bank or credit accounts, review the sign-up screens carefully and make sure that you are selecting or deselecting the permission that they will ask you for regarding their ability to track and use your personal information.

The offers that will come to you over the next 3 to 5 years will be extraordinary. For example: a retail cashier is likely to offer you a big discount on the purchase you are about to make if you "just tap your phone right here," as he points to part of his cash register. If you do that, rest assured that somewhere, on some computer, on some faraway network, a new account was opened just for you and that account will look for as much information about your shopping patterns and behaviors as possible.

The other side of this discussion is that you can, actually, get better discounts for things that you want if you are willing to trade some of your privacy. For example, have you ever received those blue oversized coupons that are sent out routinely by Bed Bath and Beyond? Did you notice that after you used every one of the $10–off coupons that they sent to your home, the amount eventually changed to $20–off. Why did they do that? Because they noticed that when you use a coupon, the average amount of your transaction quadruples. They look at the profit each time the dollar value of your transaction went up, and wondered if you would spend more money still if they doubled your coupon amount from $10 to $20. So they tested that theory by sending you a $20–off coupon, and then looked at the amount of your transaction when you used it. If they liked the result, then you keep receiving the $20–off coupon periodically; If not, you start receiving the $10 coupon again.

The point is, if you do not want to be in Bed Bath and Beyond's database, don't use their coupons. If you do not want your local grocery store to know anything about you, do not join their club program, but you will likely give up the benefit of some sale prices (which require you to use their club card to get the discount) plus the benefit of using the points you accumulate to purchase whatever you can through their catalogs.

So, it really is a cost versus-benefit-decision. If you don't use a coupon, you don't get the discount but you also don't get tracked on their network database. If you don't sign up for club card, again you lose the discounts and benefits, but you avoid getting added to another network database.

CONSUMER TIP

Pay attention to the privacy policy of every SmartPhone Wallet service you use.
If the policy is unclear: **ask**.
If the policy is unacceptable: **pass**.

If you were not aware of these issues as you signed up for all of these various programs over the years, don't worry. The next decade will give you a chance to start over on all of them.

As you use your SmartPhone, if you are offered a program that does not seem to give you enough privacy options, then simply do not join it.

There will be other programs that will give you access to the same discounts with better privacy; you just have to find them. Keep an eye out for the *SmartPhone Wallet: A Consumer Guide to the Services of Today and Tomorrow* that can point you to the best and worst of the programs available. Also, you may need to show a little patience. The best programs with the most privacy flexibility and the highest level of rewards may take some time to materialize because not all of the necessary technology is in place yet.

The Power and Convenience

Having detailed the privacy issues, we have explored the downside of this technology. Obviously, there needs to be some upside for you to consider removing any element of your privacy. The upside is two-fold:

Power through functionality, and convenience.

Here is an example.

Think of an ATM machine. The computers and databases that manage your local ATM machine are many miles away within the walls of your bank's complex processing system. But the security level of that ATM machine is so high, and the communications network that it uses to connect to the bank's network is so secure, that the machine is able to distribute cash which is obviously an activity that needs to be done right the first time because it cannot be undone once complete.

Imagine if your SmartPhone could be as secure on any network as that ATM machine on your bank's network. You could move your cash directly from your account to any person or business at will.

If your SmartPhone could be trusted by your credit card company's network you could literally use it to interact with your credit account to do any kind of transaction you liked at a cash register.

> **Imagine if your SmartPhone could be as secure on any network as your local ATM machine is on your bank's network.**

If your SmartPhone could be trusted by an alternative payment company, such as PayPal, you could initiate any type of transaction in a way that is even more secure than using your laptop.

If your SmartPhone could be trusted by the loyalty program of your grocery store, (and your airline, and your local bank, and your favorite restaurant, etc.) then your phone may be able to tell you exactly how to pay for something in a way that will earn you the most points.

If your SmartPhone could be trusted by your state's Department of Motor Vehicles, then your SmartPhone would be a more secure identification method then any two-dimensional drivers license.

You get the idea; when the SmartPhone truly becomes a "trusted device" on any network, the potential for convenience is extraordinary.

Many Choices, and More to Come

There are literally dozens of companies in the emerging SmartPhone Wallet Industry: Payment companies, loyalty companies, coupon companies, and other companies that handle processing. On the surface, the breadth of choices may seem staggering, so there is the genuine concern that all of these different products from all of these different companies will result in a form of "consumer paralysis;" which is the phenomenon of inaction caused by a lack of clear choices. In today's market, for example, you could choose to use the product of Bling, Obopay, Boku, and others -- many others.

Do you really have time to compare the different offerings from different companies? Do you have any framework for deciding which of these various products fit into what you consider to be the most optimal product for you in the future? Well, given the generally low adoption of these companies to date, it would seem that you are not alone.

To begin to categorize all of these different products, services, and startup companies, it is important to realize that they can all be sorted into one of two formats for processing payments and managing money: the Current Format and the New Format.

The Current Payment Format

Every time you use a V/MC/AMEX card, you are using a payment system – or a complex system with many participants that are responsible for moving your money, at your request, to another party (like a retail store) at the time that you want to buy something. All the companies involved in issuing your credit card to you (Issuers), and the companies that supply the credit card machines for retailers (Acquirers), and of course the ubiquitous companies whose logo appears on virtually any card that you use for payment: Visa, MasterCard, American Express, Discover/diners club. In addition, there are a host of other companies that you have probably never heard of who support these systems worldwide, such as First Data, Fidelity Information Systems, TSYS, Heartland, and a host of other companies known as "Processors." These companies operate behind the scene and manage large, secure data networks that allow your transactions to be completed.

Many, if not most of the new products and services enabling The SmartPhone Wallet are actually using the current payment format. In fact, the aim of many of these products is to simply replace the action of swiping a credit card with the action of tapping your SmartPhone.

As you can imagine, the companies that make money from the current format are highly motivated to keep you using their systems.

The New Formats

Logically, every system of payment that does not involve the current format, by definition, involves a "new" format. These are the companies that will allow you, for example, to make a purchase that is charged to your mobile phone bill instead of your credit card statement. Or, companies like PayPal that let you load your money into your virtual account and manage that money with more freedom and flexibility than you could out of a traditional checking or savings account.

And there are many other new ideas and products that focus specifically on a new format of completing payments using your SmartPhone.

So, these two formats are already beginning to conflict with one another. Before most of us have even started using our

SmartPhones to pay for things, the players are lining up along each of these two formats, and they will spend a lot of money to make sure that they get to participate in the industry, and the other format will not get to participate.

Sounds a little bit like the start of the war, right? Well it is. And one thing that is true about all wars is that there are casualties. If you remember nothing else from this book, remember that we, as consumers, need to make sure that the best products on the market do not become senseless casualties in a format war.

CHAPTER 3:

THE FORMAT WAR

Mobile Commerce companies will present you with offers that are easy to join, and easy to use, but they will make it hard for you to switch to another companies' product once you have joined.

It is called a "Format War." Two or more companies with different products compete for your attention. Of course, as a consumer, you get to choose the product you want. If you don't like it, certainly you are free to switch to another product. However, since companies know this, they can make it incredibly cumbersome for you to do so.

Now, just for a moment, look past all of the predictions that credit cards will disappear in a matter of months, and that the global economy will be changed forever, and that your whole wallet will become obsolete all because of the new technology called SmartPhone.

Here's why it's important to take a moment and do this:

The next three to five years will determine the type of technology that most of us will end up using for at least the next decade.

Why?

CONSUMER TIP

Be willing to try different SmartPhone Wallet services as they become available, and do not get hooked on the first service you try - it may not be the best.

Because one of the biggest components in a format war is the price of switching. If you make a choice to purchase one format versus the other, and it costs a lot of money when/if you realize you made a mistake, you're more likely to live with your mistake rather than incur a new cost associated with switching. The risk, of course, is that if the other format is truly superior and it does not get the support it needs to survive then that format may go away, and we all lose the advantages it offered.

HD DVD vs. Blu Ray

Take the recent format war between Blu-Ray and HD DVD. Did you kill the HD DVD? All you had to do was buy a Blu-Ray Disc player. Chances are, once you chose Blu-Ray, you never gave the HD DVD format a second thought. If you were a Blu-Ray buyer/HD killer, then job well done. You chose the better product. The Blu-Ray had 66% more capacity than HD DVD with the theoretical limit of 200 GB compared to HD DVDs theoretical limit of only 60 GB. The Blu-Ray could even record and playback at the same time off of its rewritable disc; the HD DVD could not. The Blu-Ray was not first to market, and was more expensive than HD DVD, but it was hands-down a better product. It was only in 2008 when the hefty backers of HD DVD - Toshiba, NEC, Sanyo, Microsoft, NBC Universal, Viacom – admitted defeat and declared the Blu-Ray standard the winner. That's when the consumer choice finally became clear. However, during this nine-year format war, the misinformation circulating about both formats was significant. There were outdated reports, misleading marketing, and misinformed salespeople. Also, there was no shortage of bloggers, commentators, and pundits all actively voicing their opinions on each format (some opinions based on fact, and others based on something else)[i].

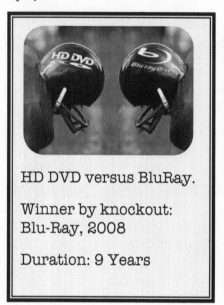

HD DVD versus BluRay.

Winner by knockout: Blu-Ray, 2008

Duration: 9 Years

Betamax versus VHS:

If you have as many gray hairs as I do, then, you remember the Betamax. Soon after they made the Betamax, Panasonic made the VHS, and they had different strategies. How many of you remember using a Betamax? Probably very few of you. That's because of a few very poor tactical moves that occurred in a short two or three year period, that allowed

VHS to win the battle and maintain their lead long enough to drown out and eventually completely kill the Betamax.

The format war between Sony's Betamax and the VHS standard introduced by JVC really only lasted about four years. For those of us who lived through that period, it certainly seemed like a substantially longer period of time.

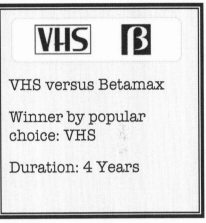

VHS versus Betamax

Winner by popular choice: VHS

Duration: 4 Years

Originally introduced in 1975, Sony was actually working closely with JVC to try to get JVC to support the Betamax format in exchange for mutually sharing the patents on the Betamax technology. Sony actually believed that they would gain JVC's support, until October 1976 when JVC introduced their competing videotape format called VHS.

Did the better format win? With the benefit of 20 years of hindsight, it appears the products were actually very much alike. They had similar features, similar capacity, and similar quality. But that was not the rhetoric of the day.

At the time, most experts considered the VHS format inferior to the Betamax, and said so publicly. Although no data exists on this, I would imagine all of the diehard Betamax fans of the day were also Betamax owners who wanted the format they selected to win because that would mean more movies were made in the Betamax format, and ultimately the Betamax user would have more choices. But, like other format wars, the consumer was largely left in the middle and for a few years many people stayed out of this new video recorder market while they waited for one of the formats to beat the other.

Despite a strong effort by Sony in 1977, in which they partnered with Toshiba, Sanyo, and Zenith, all of whom agreed to produce the Betamax product with their respective names on the product, it was not enough. By 1980, VHS had 70% of the new market and the Betamax was all but finished in terms of a mainstream consumer product. It wasn't until 2002 that Sony finally retired the entire Betamax line, which had only been kept alive over its last two decades with the professional–grade Betamax models and a new version that was popular in Japan called the "Extended Definition Betamax".

Who won and who lost? The peak of this war really only lasted four years, which was less than half the time of the HD DVD versus Blu-Ray disc format war. As the 1980s progressed, the only choice a consumer was confronted with was which brand of VHS they wanted to buy, and not whether or not to buy a VHS or Betamax. In the end, the customer was not denied a better product, since both products really performed similarly; and the length of the battle was mercifully short. However, those consumers that purchased a Betamax during those four years certainly didn't see this upside as their technology became obsolete and they had to purchase a VHS anyway.

Apple versus Microsoft[ii] Part I

Although the competition between Apple and Microsoft continues as of 2010, the format war that took place between these two companies around the time of their origin and through the 1980's is a classic example of something consumers generally want to avoid: An inferior product beats a superior product so badly in the marketplace that the superior product vanishes entirely from the marketplace. Obviously, Apple did not disappear from the marketplace, but it came awfully close. Microsoft started its operations focusing on software, and eventually began to develop the complex "disk operating system" also known as DOS. Apple, on the other hand, began its operations with computer hardware before developing its groundbreaking operating system that at the time was known as the Apple Lisa. The main innovation with Apple's Lisa operating system was the ability to use a pointing device (which we now call a mouse) to move a cursor (which we still call a cursor) and click a button on the mouse to select a specific function– this was also known as "point and click."

Apple versus Microsoft

Winner by Knockout: Microsoft

Duration: 4 Years

After developing Lisa, Apple set out to create a new personal computer. Initially, this device was internally nicknamed SAND for "Steve's Amazing New Device." Another name for this device was the now famous "Macintosh".

Apple knew that it needed productivity software to allow the Macintosh to compete in corporate markets, a sector which represented the majority of personal computer sales at the time. To do that, Apple partnered with a competent little software company called Microsoft.

At the time, Microsoft was working on its own operating system, but soon after working with Apple developing software for the Macintosh, Microsoft realized that the Apple operating system was far superior to anything that Microsoft had, or was likely to have in the near future. So, using what they learned from Apple, Microsoft began to develop a new operating system called Windows.

After the public introduction of Apple's Macintosh in 1985, Bill Gates contacted Apple's CEO, John Scully, in an effort to allow Microsoft to license Apple's new operating system both to Microsoft and two other companies, ostensibly to attempt to create a new standard in personal computing. Unfortunately, Bill Gates forgot to mention to John Scully that Microsoft had used its early access to Apple's Macintosh technology, and Apple's Lisa operating system with its point-and-click functions.

Here is where Apple was both right and wrong at the same time. Scully rejected the request from Bill Gates with the assertion from the Macintosh product manager, Jean–Louise Gassée, that the Macintosh was more than a generation ahead of the rest of the personal computer market from both a graphics standpoint, and a user interface standpoint, which was without question an accurate statement at the time. However, both Scully and Gassée believed their products to be so much better than other personal computer products, that the Macintosh would not have any serious competition in the foreseeable future, and it could rely on the higher-margin hardware sales that it was enjoying at the time.

At first, Apple appeared to be right. Microsoft released Windows 1.0, which appeared to not be a threat. However, some of the features were obviously "borrowed" by Microsoft from its experience over the prior few years developing software for the Macintosh. Including: a menubar at the top of the screen that was nearly identical to the one found on the Macintosh. Also, the Windows operating system included a word processing program called Write, and Paint that could perform simple drawing functions. The Macintosh, of course, also included strikingly similar products.

Then, Microsoft released Windows 2.0. Apple immediately recognized it as a significant improvement over Windows version 1.0, which included a host of additional features that were exactly like the Macintosh, such as: overlapping windows, multitasking, and the use of icons to identify folders, programs, and other functions within the computer. In addition to a much better operating system, Microsoft also wrote two important new programs for its window 2.0 system: Word and Excel. These two immensely popular productivity software applications are still in wide use today. Finally, while Apple continued to hold the development of its software very closely, and continued to focus on the sale of hardware as its main profit driver, Microsoft had simultaneously convinced other major software companies of the time to write their software on the Windows 2.0 format. These companies included Corel, Microtek, and Aldus.

Of course, a famous lawsuit followed in which Apple sued Microsoft for breach of its licensing agreement. But, on July 25, 1989 Judge W. Schwarzer ruled that, of the 189 claims of infringement by Apple, 179 claims were ruled as allowed by Microsoft within the licensing agreement that Microsoft signed with Apple. Also, the remaining 10 disputed items were classified as not infringements on the licensing agreement.

And that was the end of Apple's Macintosh. The Macintosh never achieved 5% of the market share, even after the explosion in the personal computer market in 1987 when the market doubled in size in just two years. And, by 1989, Windows–based virtual computers exceeded 80% of the market, while the Macintosh still floundered at roughly the same number of annual unit sales, which by this time represented only 2% or 3% of the market. It was not until Apple introduced the iMac almost 10 years later in 1998 that Apple began to return to profitability.

> ## CONSUMER TIP
>
> Remember that the company that produced the iPhone, the iMac, and the iPod could have died in the 1980's because not enough of us chose their products. The best SmartPhone Wallet companies of the future will be fighting for survival over the next five years. Don't let the good ones die!

Windows had won the first-round of the battle with its Windows 2.0 product built on the innovation of the Apple team. Think of what would have happened if Apple did not survive the difficult years between 1989 and 1998; there would be a host of products that would either be absent from the marketplace today, like the iPod, the iPhone, the iMac, and other popular Apple products all because we, the consumer – you and I – chose the Windows based format as instead of the Apple format.

At the time, nobody could have anticipated the contribution that Apple would make in later years, or, how the Microsoft Windows 2.0 malfunctions and errors would continue for another 20 years. The marketplace was convinced, and hopeful because they had already spent their money, that Microsoft would figure out how to stabilize its operating system before the next version of windows appeared.

If you purchased a computer in the late 80s, would you have chosen a Windows–based machine if you thought it would lead to the demise of one of the most innovative companies in history? Had you known about the superiority of Apple's modern products, would you have supported them in their infancy?

The answer to those questions is up to you, but the message is loud and clear. This is the choice you have right now for the SmartPhone wallet.

If you choose inferior products, and use services based on marketing campaigns, or what is easiest, and *not* through quality, you could be giving up a future line of products, money management tools, security features, convenience, and power that none of us can imagine today. As you hear more and more about the social and economic impact of new products that will be introduced over the next 20 years, remember this: these advancements really are possible, but they will be based on the market success of the products that you are introduced to over the next three–to-five years.

The War Between the Formats

As the saying goes: Those who do not know their history are bound to repeat it. Mobile payment and mobile marketing companies have always been laser focused on offering you a service that is: easy to join, and free to join, but over time collects more data and more preferences and more patterns from you until it becomes an indispensible service to you. That is when you become less likely to switch to another company's service if the opportunity presents itself.

If you are flexible, and willing to try different solutions, then you do have more than one choice. But, as the previous chapter described, most of us are not are not willing to try another solution –even if it may be better -- if the time or expense of switching is too high.

As of the end of 2010, there are over 100 different companies with products that allow you to use your SmartPhone to either buy products, or get discounts at a retail store. Some are efficient and some are not. Some are based on very old technology and others rely on modern efficient equipment. Some will simply cost you more money to use over time than others. The problem is, it is difficult to tell the difference. And, once you have made your selection, you are less likely to switch to another product.

The purpose of this book is not to attempt to dissect each company, its position and its place in the mobile eco-system. Instead, this book focuses on the big picture.

So, in an attempt to oversimplify the context, think of it this way: SmartPhone technology will either be the next evolution of the payment system as we know it today, or it will change the payment system as we know it today.

Every product in the market fits into one of these two camps.

The only question is which camp will provide the best services for you?

If enough people choose the most expensive solutions, from the biggest brands with the most marketing dollars to spend, then the new more efficient solutions may no longer be available.

Choose Wisely

You are about to change a big part of our economic system. Yes, you. How will you do that? With a vote.

This vote, however, does not have the luxury of being part of the political system. What I mean is this. When you cast a ballot for an elected official, the ballot box, or the voting booth, are dead giveaways that you are voting, right? But, when you change our economic system soon, none of these companies are going to tell you that you are casting a ballot. And certainly, none of these companies will remind you that, like most things you vote for, you may have to live with your decision for a long time.

One of the biggest lessons learned in all the "Format Wars" described is that it caused many people to choose option "C". Meaning, they didn't choose at all. Millions of consumers chose to wait it out for one of the players to emerge victorious before they made an investment or commitment.

If mobile commerce becomes a format war between the traditional credit card system versus the new breed of payment companies, then many (if not most) people will be wary of both choices and much less likely to participate in mobile commerce at all. That means we all have to wait longer for the benefits that mobile commerce has to offer.

CONSUMER TIP

The best way to decide which format is better for you is to try different products in both formats, and remain willing to switch products over time.

CHAPTER 4:

EVOLUTION OF PLASTIC

P lastic.

When talking about the subject of money, 'plastic' means only one thing: cards. Credit cards, debit cards, prepaid cards, loyalty cards. Americans tend to have wallets full of Plastic Cards. And you are probably comfortable that all of these cards perform their respective tasks when you present them at a retail store. You know how to swipe your credit card. The cashier knows how to process your "club" or loyalty card. It all works.

Visa, MasterCard and American Express (a.k.a. AMEX) are the three dominant brands in the payment system that enables you to use the credit cards and debit cards in your wallet today. The system is complex, but mature, and a variety of different types of companies are in place today that both support the marquee brands and the financial institutions that ultimately supply the actual accounts.

On one side of the system, there are "consumers." Consumers want credit cards so they go to businesses called "issuers." An issuer is typically a bank, credit union, or other company like Capital One (before they started buying traditional banks). Of course, it costs money to issue your card and manage your account, and most of these institutions also like to make a profit (credit unions do not make a profit), so issuers charge you interest and fees to make money.

On the other side of the system, you have "merchants." These are the restaurants, stores, etc., that want to accept your credit card as a payment for their goods and services. To do that, these merchants go to companies called "acquirers" which typically provide a device or software for merchants to accept your credit card as payment. As part of the acquirers service they also do some of the back-end processing that delivers an

Recognizable logos of Visa and MasterCard

approval code to the merchant at the time of purchase; to ensure that your card is good and to guarantee your bank will pay.

In the middle of issuers and acquirers lives the heart of the payment system. The true core is where Visa and MasterCard's' computers live. The reason for this is that every acquirer does NOT want to have a direct relationship with every issuer, so both camps use Visa or MasterCard, etc. as the intermediary.

The Scheme (Yes, it is Really Called That!)

The first credit card was invented in 1958. At the time, carving numbers into a wallet-sized piece of plastic was the most technologically advanced option available.

When Bank of America came out with its BankAmericard in 1958, it had a simple system in mind: a system is to move money from a consumer's account to a merchant's account on the same day that the consumer makes a purchase. That's all.

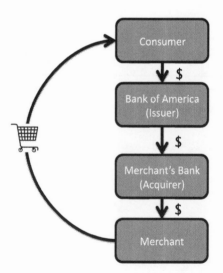

Every retail store (or "Merchant") has a bank account, some of those bank accounts are with Bank of America and the rest are with other banks. Bank of America's process was to take all of the money spent by its customers, using its credit card, and simply deposit the appropriate amount in the account of each retailer's bank account at the end of each day. This was referred to as the "four party system." The four parties were:

Consumer: a person using a non—cash payment method to purchase goods or services.

Issuing Bank: the financial institution which holds the account that the consumer will use to purchase goods or services. The first bank to do this was Bank of America, and for a while they were alone in this function.

Merchant: the retailer that wants to sell its goods or services to the consumer and is willing to accept a non-cash payment method.

Acquiring Bank: the bank that holds the merchant's account. This is where the funds from the transaction will be deposited at the end of this process.

In other words: you have a bank account, the retailer has a bank account, and you both want to move money from your account to the retailer's account in exchange for the goods and services that the retailer has to offer.

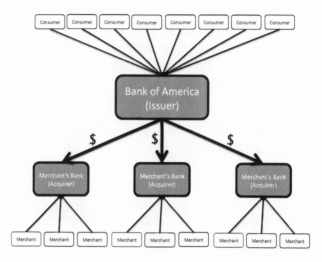

The original system was simple. At the end of each day, Bank of America would add up all of the purchases made by all of its customers, then it would figure out the sum totals that had to be paid to each retailer, and it would send the appropriate funds as a lump sum to the bank of each retailer, along with instructions about which accounts needed to be paid.

Bank of America was fond of this model because, as the above chart illustrates, all of the money flowed through their hands, which the ideal function of any bank. If Bank of America remained the only bank that issued credit cards, then the system of transferring the money would have remained easy. But, of course, every other bank wanted to compete by issuing its own credit card to consumers.

It did not take long before all of the banks involved realized that they needed a common standard in a common system to settle all of the funds within all of the accounts every day. Otherwise, if every bank simply sent funds to every other bank every day, then the result would have been significant redundancy, wasted expense, and an impossible system to audit.

This is an image of the system that would have created all of this waste:

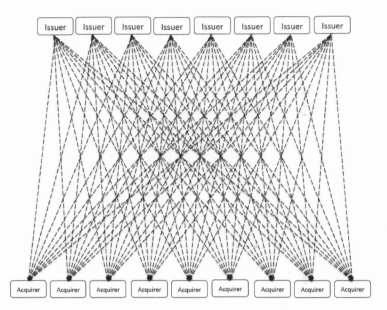

In the late 1960's, there were a handful of other banks attempting to create similar products, including First National City Bank who operated a proprietary credit card called the "Everything Card." The market was on its way to the chaos described in the chart above.

Fortunately, sobriety and common sense reigned on the west coast of the United States. A group of banks in California including: United California Bank, Wells Fargo, Crocker

National Bank, and the Bank of California, formed the Interbank Card Association (ICA). The ICA implemented a product called "MasterCharge", which, of course, is called "MasterCard" today. To operate the MasterCharge product as one system, they created a new kind of payment "Scheme."

Enter the fifth party to the four-party system: The Scheme

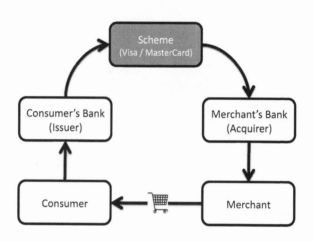

The Scheme is essentially two sets of rules. The first set of rules tells all of the parties involved how to determine if a transaction request is approved or not. The second set of rules determines how money is transferred at the end of each day to settle all of the accounts at each respective bank. The Scheme does **not** actually handle any money, but that element will be addressed later in the Chapter.

The MasterCharge product eventually forced Bank of America to give up sole control over its BankAmericard product and Bank of America started licensing its product to other banks outside of California in 1965, until it finally gave up control by creating National BankAmericard, Inc. in 1970. This was a non-profit cooperative that was owned and operated by the banks that had licensed the original BankAmericard product. As you probably know, National BankAmericard changed their name to Visa.

Interestingly, Visa and MasterCharge did not have to make a profit. They were set up to make enough money to operate the service that they provided to the banks. But, at the time, banks wanted to make this product as attractive as possible to merchants because ultimately the merchants had to pay for these services and buy the physical equipment to let the customers use these services in their stores.

Over time, more parties joined the system when the banks realized that managing the flow of data was both unprofitable and was a long-term skill that most banks found inefficient, and better left to a specialist. Enter the Processors.

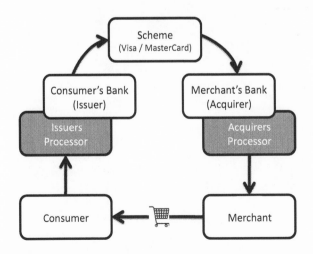

While banks and credit unions are regulated entities that are responsible for carefully handling money, the "processors" are responsible for handling information and instructions.

Have you ever wondered why only a few seconds of time elapses between the time you swipe your credit card, and the time that the cashier receives an approval code to complete your purchase transaction? It is the speed of the processors that make this possible.

Neither Visa nor MasterCard handle any money.

Instead, they create and transmit a complex set of instructions to move money – lots of money – between banks, and they contract with large banking institutions (called "settlement banks").

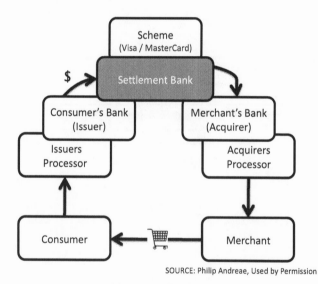

SOURCE: Philip Andreae, Used by Permission

So, the Scheme is responsible for sending the transfer instructions, and the Settlement Banks are responsible for ensuring that all of the funds are transferred accurately between all of the banks involved. The Scheme players (namely Visa and MasterCard) have a contract with specific banks to handle this function that enables the Scheme to handle the communication but avoid actually handling money.

Transaction Approval

As described earlier in this Chapter, the Scheme is responsible for two sets of rules. One of the rules describes how money is ultimately transferred to the merchant's bank account. The other set of rules is regarding the "transaction approval process". Now that all the players in the Payment System have been identified, and we have discussed how funds are moved through the system once a day to settle all the transactions of that day, it is much easier to describe how the transaction approval process happens.

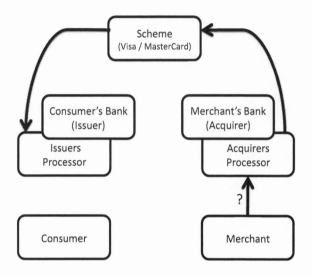

Essentially, when you give the cashier a credit card to pay for the goods you want to buy, the cashier starts a process that asks the question, "Will you promise to pay." Of course, the cashier is interested in your answer to that question, and will eventually ask you to sign your name, or enter your PIN. But, **really,** they want an answer from your bank. So, that question is communicated using most of the same participants in the system we have discussed. The chart above shows how the approval is requested, and the reply simply comes back to the merchant through the same channel in the other direction.

What is important to note about this approval process is that information is routed through a **communication network**, and the answer is sent back through that same **communication network**. This will become more important later in this Chapter when we discuss how the Mobile Network operators, which tend to be skilled at sending information back and forth, have an interest in participating in the system.

Payment System Profit

As discussed earlier in this Chapter, the "Schemes" were originally not–for–profit collectives. Eventually, that changed also, and today both MasterCard (the successor of MasterCharge) and Visa are both for–profit, publicly traded, and highly profitable corporations.

So, a system of payment that began as a simple way to get money from a consumer's accounts to a merchant account now supports six separate profitable types of companies and institutions: the Issuing Bank, the Acquiring Bank (most banks can perform both functions, by the way), the Issuer's Processor, the Acquirer's Processor (again, most processors can also perform both functions), and the Schemes.

Note that Credit Unions can be Issuers and Credit Unions are not–for–profit organizations. So, other than Credit Unions, all of these other types of companies are in it for the profit.

Where does all of that profit come from? You. The consumer.

Actually, the merchant is charged for the fees associated with processing a payment, but the common practice around the world is for the merchant to inflate their prices enough to cover the expense of this payment system.

Essentially, the merchant pays two different types of fees: an Interchange fee and a Transaction fee[iii]. The Interchange fee represents a percentage of the transaction amount, and the Transaction fee is typically a fixed amount of money that the merchant pays regardless of the amount of the transaction. The sum of both the Interchange and the Transaction fee typically cost the merchant between 1% and 4% of the transaction amount, depending on the merchant's agreement with their Acquirer, the amount of the transaction, and the payment type. Larger merchants (like Wal-Mart) generally negotiate lower fees that smaller merchants (like a local bookstore that is not part of a franchise or a national chain).

And, the revenue from these fees is significant.

Since going public in 2008, Visa's 1.8 billion cards made 66 billion retail transactions worth $4.8 trillion worldwide last year. In June 2010, Visa owned 57% of the combined U.S. credit and debit card markets, compared with 53% in 2007. MasterCard had 25%. Operating revenues grew 10% to $6.9 billion in Visa's 2009 fiscal year. Net income was $2.4 billion.

What Payment Companies Want

With all of that profit going around, the Payment Companies want everything to stay exactly as it is today. But, they are being forced to address this new concept of the SmartPhone Wallet, so they are running tests to see if they can capture this market, too – and quickly.

V/MC/AMEX and all of the other participants in the payment system as described in this Chapter logically want the status quo, and they are not alone. The payments system is mature, widely regulated, wholly adopted, and popular. Consumers know where to go to get a credit or debit card, and merchants know where to go to have those cards accepted. Banks and credit unions know how to interact with V/MC/AMEX and so do the merchants. So, as SmartPhones are beginning to appear to be another way to bank, and another way to pay, V/MC/AMEX are highly interested in keeping the systems used for credit cards in control of the new SmartPhone payment systems also. So, V/MC/AMEX found some partners and launched a test in Sitges, Spain to try to figure out this new market.

The Sitges Test

La Caixa Bank, Telefonica and Visa teamed up to launch a mobile payment system market test in Sitges earlier this year. The test involves 1,500 people and roughly 500 retailers and restaurants. The system being tested incorporates Near Field Communication (NFC) technology in Samsung phones. The consumers in this test market were told to simply move their SmartPhone near the device where they once swiped their credit card at a cash register. However, in this test, they do not need to use any kind of card, they ONLY need their SmartPhone.

The project's objective is to study the future widespread implementation of shopping using mobile phones. Due to the relatively large number of participants, the project is the most important experiment of its kind to take place with Visa and its merchants to date.

Press conference announcing the Sitges Test

In Visa's market test in Sitges, the payments are restricted to transactions at retailers, but the future system will also allow customers to use their SmartPhones for other purposes such as buying a ticket on public transport, or sending money to a friend using only their SmartPhone.

Visa's system is based on the SIM card that is installed in every GSM phone (see Chapter 1: Glossary) In this case, the SIM cards contain an additional code (other than the code that enables the phone to connect to the GSM network), which triggers a transaction similar to a standard debit or credit card. In fact, the customer can even select which credit or debit provider they would like to use for the transaction.

To participate in the test, 1,500 people in Spain have been provided exclusive Samsung mobile phones that include the NFC technology. The people have also been given some minimal training since this product is not yet in the collective unconscious of us all. All participants met the requirements of being customers of La Caixa, having a Movistar mobile phone account and a valid Visa card. Then, they are set loose to see what they buy, how often, and (of course) how much money they spend.

While information from Visa is still slim as of the writing of this book, this test-run appears to be a good example of a single application running on a SmartPhone that is capable of selecting multiple payment methods. Quite possibly, this is the first real test of this type of SmartPhone Wallet. So, like every other deployment, security is the top issue.

Kim Faura, Director General of Telefonica in Catalunya, says that the Visa system being tested in Sitges has "the same security as a card," which is the main selling point of the program according to William Gajda, the Head of Mobile Innovation for Visa. Mr. Gajda is the man responsible for courting SmartPhone Wallet startup companies into using Visa's current format instead of a new format.

Essentially, all this test does is let the consumer trade their plastic cards for their SmartPhone. Clearly, if an entirely new payment Scheme is possible by leveraging the technology of the SmartPhone and the modern Mobile Networks, Visa does not want to find out.

What Mobile Network Operators Want

The current (old) payment system works just fine; why would anyone consider changing any of this?

> **Mobile Network Operators want to know if a more efficient payment system is possible.**

Because there may be a better, more secure, and less expensive way!

The worldwide payment system that handles credit cards and debit cards is responsible for about $11 trillion dollars a year. Could the mobile network operators handle that?

The answer is: they don't have to.

Again, Visa and MasterCard do not handle any money; only banks do that. So, these companies keep contact with Settlement banks to ensure the money moves from bank to bank based on the instructions of Visa and MasterCard. This raises three critical questions:

If Visa and MasterCard can hire settlement banks to handle the flow of funds daily, why can't another company - such as a mobile network operator (like Isis), or an alternate payment channel (like PayPal) -do the same thing?

Could a more efficient system, with fewer participants, exist?

Would such a system present cost savings, or consumer convenience?

The companies making money in this current payment industry do not want you to find out, but the Mobile Network Operators do.

Verizon, AT&T, and other mobile network operators, want to consider something new. They think it could be an interesting idea to cut out V/MC/AMEX and their payment peers from making any money on transactions by simply charging your purchase to your cell phone account. If such a system costs less to operate, it could lead to lower fees to the merchants who could pass that savings on to you.

Granted, the current format is formidable, and only the truly arrogant would imagine building a new startup designed to displace mature worldwide brand names like V/MC/AMEX. After all, they are well-established, worldwide brands; how could they possibly be vulnerable?

Here's how.

At its core, the payment system is simply a huge communication system. It is a secure, standardized, regulated, and extensive communication system; but it is just a communication system.

The core function of V/MC/AMEX is about as complicated as completing a phone call across two-networks, but it is more secure. So, what worries V/MC/AMEX is that mobile network operators like Verizon Wireless and AT&T Wireless will adopt the necessary security standards, comply with banking regulations, and then take over the trusted brands that we all know so well.

In other words, Verizon looked at V/MC/AMEX and said, "I can do that."

A Mobile Approval Process?

Here is one way that a mobile network operator could make this happen. What if the electronic cash register sent a message to the mobile phone (assume that is easy to do for the moment)? That information could be sent directly to the Issuing bank for approval and the Approve/Decline message could be sent through the same channel in reverse. The merchant may need to inform its Acquirer how much money to expect from the Issuer(s), but otherwise the information flow would look like this:

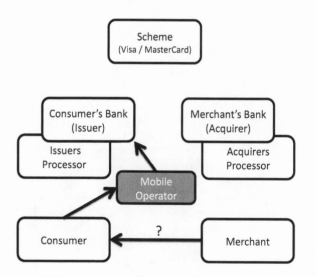

That method would include only the mobile network operator and the consumer's bank that issued the credit or debit account, which would leave out the rest of the participants in the current payment system.

A Mobile Payment (Settlement) Process?

Earlier in this Chapter it was explained that the Schemes do not handle money; opting instead to issue a complex set of transfer instructions through an intermediary -- the settlement bank.

Is it possible that the mobile network operator could also contract with a settlement bank to accomplish the same thing? If that happened, could the mobile network operator act as the processor for the issuers and the acquirers?

Could the transfer instructions be managed solely by the mobile network operator and the settlement bank? Would such a system cost less to operate and manage than the current format of the payment system?

Here's what this system could look like.

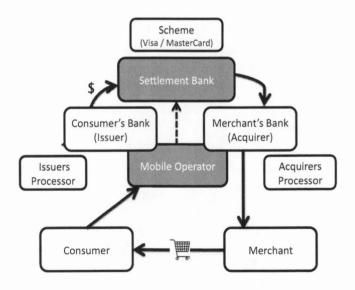

Want to go further? It is possible.

The mobile network operator could send the transaction directly to the bank and trigger a transaction (known as an ACH transaction) directly from your bank account to the merchant's bank account.

Here's how much simpler the system could look:

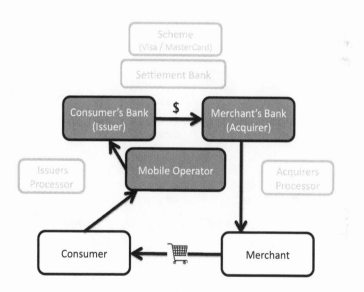

You would have to choose "ACH" as your payment option, and your SmartPhone would need to be certified by your bank (discussed in the next section), but it would work.

Further still?

What if the Mobile Network Operator just sent the transaction to PayPal?

After all, better than an instant promise to get paid is the instant payment itself. In this scenario the merchant's PayPal account would instantly show the receipt of funds. Of course, the consumer would need to choose PayPal as the payment option, and the Merchant would need to have set up a PayPal account also, but its possible.

Here's what this system could look like.

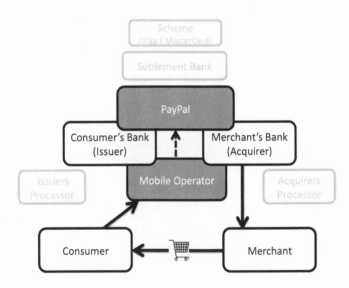

Do you see all of those light-gray boxes that have NO connection to any of the other boxes?

Each of them represents an **extraordinarily** large and wealthy group of enterprises, all of which will be willing to spend millions or billions of dollars on marketing to convince you – YOU – not to try any new Payment System Formats. Of course, an equally large amount of money will be spent by the Mobile Network Operators and other companies to promote the opposite message.

Again, here is the central point of this book – the best services should win, not the biggest advertising budget.

No one knows if a new payment system will be more expensive or less expensive; more efficient or less efficient; more convenient or less convenient -- **unless it has a chance to succeed in the marketplace.** Also, we don't know if the process flows described in this section are the more effective use of the abilities of the mobile network operators unless we give alternative methods a chance to succeed in the marketplace. Finally, we do not know if this (or any) new format for the payment system will yield advantages for the consumer or the merchant unless these options are allowed to succeed in the marketplace.

> We will never know if another payment system is more efficient and less costly unless it succeeds in the marketplace.

Mechanics Are in Place

For any new payment system to work, three things are needed:

- Hardware (SmartPhones),
- Networks (mobile phone networks); and
- Security.

Not only are all three these in existence right now, but also the technology is improving and growing fast.

SmartPhones are getting more popular, and it is happening quickly. For example, according to the Nielsen Company, SmartPhones accounted for only 10% of all mobile phones active in the US in the middle of 2008. By the middle of 2009, that figure grew significantly to 16%. One year later, in the middle of 2010, that figure had jumped to 25%. Given that there are over 270 million active phones in the United States, that equates to approximately 60,000,000 SmartPhones now active in the USA, up from about 22 million only two years ago in 2008.

GSM penetration is also growing fast. According to Telegeography, GSM has grown to an astonishing 70.5% of all of the 5.1 Billion active mobile phones on earth.

Market Size

The addressable market of companies capable of supporting the SmartPhone Wallet in the USA alone will be about $50 Billion by 2014[iv]. While this number may differ from other estimates in the marketplace, it is likely the most accurate because it incorporates potential revenue to all mobile wallet companies from all three major revenue sources that are likely from this new industry.

Here is how that could happen.

First, mobile wireless companies have the opportunity to make money by processing a portion of the transaction. After all, a flat fee charged to the merchant in exchange for functionality during the transaction is already a standard in payment processing today.

Second, as we discussed, they have the chance to make money from a percentage of the transaction: commonly referred to as interchange fees.

For a mobile wallet company to participate in interchange, they would need to take over all or most of the role from what is presently called the acquirer. This could be extremely difficult for a new party to accomplish, and would circumvent the system that is currently established. However, if the transaction could be settled through the SmartPhone, not just triggered by a SmartPhone, then this alone has the potential for significant cost savings and network efficiencies. In short, merchants are accustomed to paying approximately 2% to 3% whenever a debit card or a credit card is used. So, hypothetically, if the transaction can be processed directly through a SmartPhone, and this transaction indeed costs less to facilitate, then it is likely that the provider for the service may be able to charge lower interchange rates to the merchants.

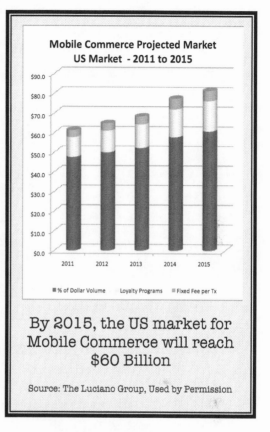

By 2015, the US market for Mobile Commerce will reach $60 Billion

Source: The Luciano Group, Used by Permission

This alone would speed-up adoption and the attractiveness of the product, perhaps offsetting any reluctance on the part of the merchant to try something new.

> "I hear all the time that we're dinosaurs and that we should just make room for the new guys." - Visa's new head of Mobile Payments, William Gajda.

The final component in this model is loyalty programs. The potential to manage multiple programs through a single service, may cost less than the current average on an annual basis, and may produce a sizable revenue stream and cause remarkable consolidation within the loyalty business for the next few years.

Could the "VISA" Brand Ever Really Go Away?

Yes, it could.

V/MC/AMEX and their peers are acknowledging a serious threat in the future: the SmartPhone. All of these companies make their money by taking a very small percentage of every transaction that uses their services. But what if you could complete a transaction without their services? What if you could walk into any store, restaurant, doctor's office, or hot dog stand and buy something without using their services? What is the alternative, and who are the potential players that could make it happen?

So, what will it take for any of these new formats to succeed in the marketplace?

Let's start with the #1 consumer concern.

Security

The main obstacle to any mobile payments adoption in the United States is the "security" issue.

According to the Harris Interactive and Billing Revolution Study: 93% of U.S. adults (93%) own a cell phone, and nearly half of these adults (45%) think it is at least somewhat safe to make a purchase through their cell phone with 26% saying they think it's fairly or very safe to do so.

Back in December 2007, Harris Interactive conducted a survey and found that 63% of cell phone users were "very concerned" about transaction security on a mobile phone.

So, within the last 18 months, sentiments have changed.

In September 2010, Gartner weighed in on this topic with an estimate that "fraud detection tools for mobile commerce are lagging. Because of the improving browser experiences on SmartPhones, mobile commerce and transaction execution are set to increase rapidly," William Clark, a Gartner analyst, said in a statement. "Enterprise applications must detect fraud in these mobile environments, but fraud detection tools available today that work in [wired] computing environments don't work well or at all in the mobile world," Clark said. Tools to detect fraud in the mobile space are in the early stages of development, and he estimated it will take until at least 2012 for them to mature.

Below is a discussion of a few of the current security measurements available and in development.

Device ID

Mobile device identification is a JavaScript running on a server that a mobile user logs into. The script captures information about a user's browser and mobile phone.

If the mobile phone user accesses the web with an application that is browser-based, the Java script application captures unique browser identification information and data to uniquely identify the mobile phone. If the application is installed on the device, the application can also gather the phone's serial number and network card number to forward to the merchant (whether the transaction is brick and mortar or eCommerce) -- but only after the user opts in to allow that data to be transmitted.

Location Information

To prevent mobile fraud, the merchant can also use the SmartPhone's location information and this only requires that the SmartPhone be turned on. Mobile phones can forward location information based on GPS data, but also require user opt-in. Locations can also be received by mobile network operators employing software tools that don't require user opt-in. Essentially, if a customer tries to initiate a transaction while standing in a retail store, the retailer would use this technology to establish whether or not the mobile phone is geographically located at the exact same longitude and latitude as the cash register. On the whole, SmartPhones tend to be excellent at security (again, when was the last time you received a fraudulent charge on your mobile phone bill?), but this method could be even more effective in combination with another security method.

The "Text Message" Confirmation

Another form of providing transaction security on mobile phones includes the use of text messaging. Also known as "SMS" or short message service. Essentially, the company that is authorizing the transaction (and there are many different services and corresponding companies in the marketplace today using this method) sends a text message to your SmartPhone and the text message contains a code. That code needs to be delivered to the cashier who is performing your transaction in order to complete the sale. Unfortunately, text messaging is an unsecure communication channel and is the subject of frequent attacks from spammers, hackers, and other scams.

The businesses that use text messaging as part of the secure authorization process during a SmartPhone wallet transaction at a retail store are, in fact, providing a relatively secure service by standards common in the marketplace at the end of 2010. In other words, if you use one of these services, you are not necessarily at risk. The reason for this is that even if someone intercepted your text message with your approval code, they would need to also have access to all of the information that the retail cash register has as well in order to clone your transaction, which would serve no benefit to the hacker, and present no loss to anyone except for the merchant who would conceivably stand the risk of getting double charged.

While there have been no such attacks known to date that take advantage of text messaging as part of the security protocol, there are other services that are significantly more secure, and have less risk of a hacker or other malicious entity figuring out a way to exploit the weaknesses in the security system. So again, this method could also be combined with another security method to increase its effectiveness.

Tapping / Dragging

Many people use the security feature of tapping their finger on the face of the SmartPhone keypad as a security function to unlock their phone, including a four-digit code that they have to enter before being able to use the phone itself. Another version of this is dragging your finger along a specific pattern displayed on the SmartPhone. Although the keypad is a useful security function, it doesn't work in the current payments system because the cashier at the retail store needs to enter the approval code, not the other way around. Also, according to PC Worldv it may be possible to determine your pass code by the oily smudges left behind on your SmartPhone. However, this system would come in handy if the mobile network operator was part of the approval process because then the customer could enter their approval code directly into their phone.

Toshiba G500
Smartphone with
fingerprint reader

Biometrics Security

The term "biometrics" means the practice of measuring some part of your body that is unique to you, such as fingerprints, retinal scans, or a picture of your face. The idea of pressing your thumb onto the screen of your SmartPhone to authorize a retail transaction is interesting, but not practical. Again, the oil that naturally rubs off of your skin and onto the face of a SmartPhone every time you touch it creates a residue that makes it difficult for the phone to accurately measure the detail of your thumb prints or fingerprint every time. But there is another way to do that: a separate device on your SmartPhone.

Trusted Service Manager (TSM)

The general idea of a "trusted service manager" is that a contactless payment initiated from a mobile phone in a secure environment is the "next logical step in the development of applications and services[vi]" to be leveraged by SmartPhone technology. The logic is that consumers would benefit both from the convenience of paying for goods and services with the new payment channel, but also because of the added security that the SmartPhone could offer by using a new system supported by a new function called the TSM."

CONSUMER TIP

You will know when your SmartPhone is being certified on a network by a TSM because it will seem like the same process that you went through when you first activated your SmartPhone, except you will be doing it at your bank (for example) and not your mobile phone store.

Two things are necessary to make this technology work. The SmartPhone has to be enabled with: an embedded NFC chip as described in the previous Chapter and a Universal Integrated Circuit Card (UICC), also known as a Smart SIM card.

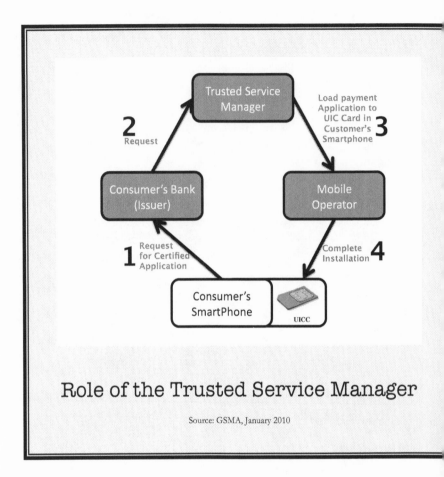

Role of the Trusted Service Manager

Source: GSMA, January 2010

The idea is that the mobile network operator uses the technology in the SIM card to ensure that the SmartPhone is secure and properly authorized. In effect, the mobile network operator manages the "trust" between its network and the SmartPhone using the SIM card inside the SmartPhone.

However, a smart UICC card is capable of ensuring the security and identity of your SmartPhone on <u>MANY</u> different networks, and not just the network of your mobile network operator. The Trusted Service Manager would be responsible for simultaneously certifying many different networks on to each SmartPhone. In the figure above, the participants are defined as follows:

Customer: subscriber of a mobile network operator and owner of a SmartPhone that is enabled with NFC and a Smart SIM card.

Merchant: retail store or other vendor that is capable of accepting a payment that is initiated from a mobile phone, not just a credit card. In addition, the merchant has to be able to accept NFC communication from a SmartPhone using the equipment at its electronic cash register.

Issuer: a financial institution, such as a bank, credit union, alternate payment system (PayPal) and any other regulated company that can hold authority and security over either a consumer's funds, or of a consumer's credit account. It is important to note that this does not necessarily need to be an account that is connected to a plastic credit card or debit card. Rather the account simply needs to exist and be compatible with the system. In other words, **no plastic required.**

Mobile Network Operator: owner of the Smart SIM card. This operator provides all of the communication between the SmartPhone and its network, or the SmartPhone and the Customer's Issuer.

Trusted Service Manager: a subcontractor of these other parties to implement the initial security authorization of the SmartPhone, and/or continue operation and ongoing management of the certificates and the authorization tools to consistently ensure that the SmartPhone is properly authorized on each network.

SECURITY TYPE	WALLET TYPE		
	Mobile Web	Handset	Application
Device ID	★	★	★
Location		★	★
Text Message	★	★★	★★
Tap / Drag	★★	★★	★★
Biometrics	★★	★★★	★★★
TSM Enabled	★★★	★★★	★★★

LEGEND: ★★★=Excellent Security   ★★=Acceptable Security   ★=Must be Combined with Another Method

Not all security types are created equal. In the chart above, notice that some of the security protocols available today are really only adequate to play a supporting role, while the TSM enabled (or certified) SmartPhone represents the highest level of network security currently available.

The next Chapter discusses the early adoption of some features of the SmartPhone Wallet that are happening around the world, and most of it is happening in surprising locations.

CHAPTER 5:

AROUND THE WORLD, TODAY

Again, the SmartPhone Wallet is an empty vessel that contains whatever accounts, coupons, payment types, and other items you like. Back in Chapter 1, we described the three general types of these "empty vessels":

1) Internet Browser on your SmartPhone;
2) The SmartPhone itself using Applications; and
3) An Application that is able to function across multiple accounts and programs.

Most of the examples that have reached significant consumer adoption levels in their respective markets are examples of #2: Individual Applications.

There are several examples worth noting of individual applications that are deployed in the marketplace. Interestingly, these examples are predominantly in developing nations.

Read further to learn more about what's out there now.

Individual Applications

One of the main components that all SmartPhone Wallets will eventually carry is the ability to access an account and use it to immediately send money, even if that account is not a bank account.

For those of you who think that the great developed nations of the world need to figure out the mobile wallet so that developing nations can benefit -- think again. Not only have some of the poorest countries in the world figured out the optimal business model for mobile payments already, but the solutions they have been using for years are perfectly suited for their customer base. Here are some examples from around the world today.

Kenya

In this market where the penetration of credit cards, debit cards, and bank accounts is low. Cash is the vehicle by which most transactions occur and most debts are paid.

Mobile payments are not only remarkably more secure than cash, but also create a modern audit trail that allow the user to dispute charges and review expenses, along with all of the other benefits of an electronic payment system.

These are the circumstances in Kenya.

So, they developed a mobile payment system called 'M-PESA,' and it is one of the most mature services in the world of mobile payments today. The service was created by Safaricom.

In most emerging markets, electronic payment options are relatively few for the average consumer. Currently in Kenya, about 10% of the population has bank accounts, and an even smaller percentage have credit or debit cards. That means this is a largely cash-based society.

How does a person pay a bill with cash in Kenya?

Typically, they get on a bus or some other form of transportation and they go to the place where the debt needs to be paid. So, when M-PESA enabled a consumer to transfer money across the country by pressing just a few buttons on their mobile phone, consumers saved a significant amount of time and effort compared to the journey that they would have had to complete in order to pay the debt in cash.

At the end of 2010, M-PESA was the most successful mobile wallet application at work in the world today. However, it operates in Kenya – a developing country – and it remains to be seen if its success can be duplicated in other markets around the world.

Its success deserves close inspection:

Within the first 40 months after it launched, M-PESA had reached 70% of the households in Kenya. In addition, and perhaps more importantly, 50% of the impoverished population who generally do not use traditional bank accounts, have opted to use the service to manage their money.

Here's how it works.

The customer deposits funds at any one of a large number of outlets across Kenya. Then, those funds are credited to their M-PESA account that is managed by the mobile network operator, Safaricom. Note that Safaricom does not act like a bank, because that is not its expertise, nor is Safaricom licensed to do so by the banking regulator of Kenya. Because of this, Safaricom deposits the actual funds in one of several commercial banks, which are carefully regulated by the banking authority in Kenya. In addition, the funds are held by a Trust and are not accessible by Safaricom, so (contrary to popular belief) the mobile operator cannot use these funds in any way; they simply do not have access.

M-PESA supports peer-to-peer fund transfers in Kenya.

Source: Safaricom, Used by Permission

Even if Safaricom went bankrupt, the creditors of Safaricom would also not be able to claim the funds deposited by consumers into the M-PESA accounts. This is a requirement from the Central Bank of Kenya that oversees M-PESA. The funds remain the property of M-PESA customers at all times.

Once the customer adds funds to their M-PESA account, that customer can transfer money easily to any other person or business that uses M-PESA also. These transactions are far more secure than cash, and can be monitored by Safaricom, which manages its own anti-money laundering system that is said to be at par with other banks and financial institutions in Kenya. Every time funds are loaded or unloaded within an account, an electronic log is created and stored by the system, and the Central Bank of Kenya gets regular reports regarding all payment service providers in country, including M-PESA transactions.

Although the combined balance of all the M-PESA accounts represents just 0.2% of bank deposits in Kenya, the total number of M-PESA transactions reached about 70% of the volume of electronic transactions in the country. These transactions are generally small, accounting for a mere 2.3% of the value of all electronic payments combined. M-PESA's success means there is a real need for small electronic transactions.

The M-PESA agents, or store owners, are essentially 'super-users' of the service, and are required to process all transactions through a Safaricom SmartPhone -- not their cash register or point-of-sale system. In effect, the agents need to keep an inventory of mobile money, just like they need to keep an inventory of other goods in their store with the purpose of selling these items to customers. So, just like inventory management of physical goods, agents must invest some of their working capital into M-PESA credits. The main difference between consumer M-PESA accounts and agent M-PESA accounts is that agents have higher transaction limits.

Training for these agents is very good. After the agent is recruited by Safaricom, a detailed due diligence process is completed on each agent. Safaricom regularly ensures that each agent complies with, and receives, specific training. Agents are regularly monitored and re-trained by Safaricom mainly to ensure that any customer anywhere in Kenya has the same experience at any M-PESA agent.

When the customer originally opens an M-PESA account, they need to present original identification (no photocopies). Agents are carefully trained to perform registration checks and can be suspended if they do not comply with the exact same procedure for all customers. In addition, the agent needs to carefully check the customers ID card any time they request a transaction.

But the security is not limited to simply checking an ID card; M-PESA uses triple–factor authentication:

1) The SIM card is authenticated;
2) The ID card is presented, and;
3) The customer must enter a PIN that is unique to their account.

Regardless of where you live, you probably do not have the same level of security when you handle cash.

Read on for another real-world example that was launched after a massive natural disaster.

Haiti

Late in the evening on January 12, 2010, a terrible 7.0 magnitude earthquake struck Haiti. The epicenter was located only 15 miles west–Southwest of Haiti's most densely populated city: Port-au-Prince. There were a total of 59 aftershocks ranging in magnitude from 4.2 to 5.9. The United Nations has estimated the death toll between 250,000 and 300,000 men, women, and children. In addition, approximately 300,000 people were seriously wounded in the area affected by the earthquake. The population of Port-au-Prince is only about 2 million people, which means around

30% of its population was either killed or wounded in this catastrophic earthquake.

In the months that followed, over $5 billion was pledged by the United Nations to support the rebuilding of Haiti, exceeding the $1 billion goal that the United Nations had set earlier.

One of the many philanthropic organizations to respond was Mercy Corp.

Mercy Corp paid it workers in Haiti directly to their mobile phones after the earthquake of 2010.

Source: Open Revolution, Used by Permission

Mercy Corp. consulted with a mobile payments company called Open Revolution to design a program to use mobile technology for the benefit of this devastated city. On September 21, 2010 Mercy Corp. - with partners Unibank and mobile network operator Voila - announced a program that would distribute funds through Mercy Corp.'s "cash–for–work" program. This is a program where Mercy Corp. paid people in devastated areas for their work during the recovery. In every previous instance of this program the funds were distributed to the workers in cash.

However, this milestone program allowed for workers to be paid directly to their SmartPhones.

After the funds were distributed to an account that was accessible by the worker's SmartPhone; the worker could either use those funds (or credits) to make purchases at grocery stores and other types of retail stores that quickly adopted this payment technology. In addition, workers could also "cash out" at a number of authorized locations, like the one shown in the picture on the previous page.

Standing at an authorized location, this is an actual photo of the text message received by one of the workers with the approval code that the worker needed to give to the cashier to complete the transaction and to distribute the cash to the worker.

The humanitarian effect was remarkable. Distribution of funds was instantaneous, as opposed to the previously arduous, and insecure process of handing out cash. Plus, once the funds were distributed they could be used to pay anyone, anywhere in the country that used the same program. Without the program, the recipients of these funds would have needed to physically travel to places in the city or surrounding territories in order to pay a debt in cash.

According to Unibank, only about half of the adult population in Haiti has a bank account. So, for those people without bank accounts, this service saved significant time and effort.

Later, on December 10, 2010 the program was re-named "T–Cash" and made available to the general population in Haiti, and rolled out nationwide.

Based on the success of the program so far, and the financial inclusion that the program has created in Haiti, the three companies delivering the service fully expect the rollout of "T–Cash" to transform financial services in the nation of Haiti.

South Africa

Vodafone took the model of Safaricom's success in Kenya, and decided to test it in a more developed country -- South Africa, under the a company name -- "Vodacom." This program is a joint venture with the incumbent South African phone-company, Telkom South Africa. Vodafone owns a majority stake in Vodacom.

This test is extremely significant because, if it works, Vodafone could potentially rollout the program to 40 other countries worldwide. They are by far the largest mobile phone operator in the world. They run most of their operations with other partners in the form of jointly owned subsidiaries, some of their assets are even branded under something other than Vodafone. In Europe, Vodafone operates in 39 countries. In the Middle East and Africa they operate in 12 countries. Vodafone also operates in 17 countries in Asia, and four countries in North and South America.

Japan

There are some success stories to tell about in the developed world also. Predictably, the Japanese were early adopters to this new technology. As early as 2002 and 2003, consumers were able to use the web browser on their SmartPhones to access websites that had merchandise, services, and other content available for purchase.

Similar to a typical online transaction from a desktop computer, this approach required that the customer enter their credit card number onto the website through the interface of their SmartPhone web browser. In 2004, NTT DoCoMo, Japan's largest mobile network operator, began selling SmartPhone handsets which used an interesting chip called FeliCa, which made it possible for a SmartPhone to both encrypt and communicate multiple forms of secure data, including bank account numbers, credit card numbers, account balances, personal identification, and other elements. The FeliCa chip communicates through a form of a radio wave to a nearby electronic cash register or vending machine, (similar to NFC[vii]) enabling the SmartPhone user to make a purchase using only their SmartPhone and an existing cash or credit account.

Essentially, consumers can download an application for the multitude of payment options that exist in Japan today, including accounts that do not require the customer to have a bank account (similar to both M-PESA in Kenya and T-Cash in Haiti), traditional credit cards, traditional bank cards, and other products. In Japan, a credit card issuer is not necessarily a financial institution. Most major retailers and manufacturing companies issue credit cards that can only be used for purchases of their products, and in some cases these accounts are never issued on a card at all - the account itself exists only on the SmartPhone.

There are 127 million men, women and children in Japan, and over 28 million registered mobile payments accounts. It could be years before the US and Europe catch-up to Japan's penetration of these SmartPhone Wallet services.

Here at Home

A host of upstarts with names that are reminiscent of the "dot-com boom" such as Bill-to-Mobile, Bling Nation, Boku,

Obopay and Zoompass think they might be able to undermine the payment system as we know it.

Their idea: If we can use the world's 5.8 billion mobile phones for mail, movies and messages, why not for money, too?

Boku, Zoompass, and Obopay are really money transfer service that enable you to send money to other mobile phones quickly and easily, and let you buy things online using the credits from the respective company. But neither can be used to buy things at a cash register (unless you use their branded MasterCard prepaid card which is automatically attached to the account). These players do not really give these out, so V/MC/AMEX do not have very much to worry about because these start-ups have relatively little ability to let the user buy anything at a retail store.

Bill-to-Mobile, however, bills your transaction directly onto your mobile phone statement along with your regular mobile charges. This is where V/MC/AMEX start to see the real threat for several reasons.

Telecom carriers – like Verizon Wireless in this case - have deep pockets, are accustomed to critical issues like security, and are experienced at managing complex customer invoices. But a visit to the Bill to Mobile web site shows that a customer can only use the service for a short list of online merchants. So what, you might say.

My answer: the model "T" Ford sold only 10,000 units in its first year.

A strong business development plan will bring in new merchants to Bill to Mobile over time. More importantly, when Bill to Mobile begins using NFC, the opportunity to use your SmartPhone at your favorite merchant becomes a reality.

What then?

You conveniently get one bill every month that has your cell phone charges, and all of your purchases. And, you no longer need any plastic in your wallet

Bling Nation is a different story. Bling already uses NFC in the form of a sticker to enable you to use your PayPal account at a cash register. This is significant because the entire transaction can be completed using no issuers, no acquirers, and no money for the payment incumbents like V/MC/AMEX.

PayPal could also prove to be a formidable opponent in the long run because of its deep pockets and breadth of more than 87 million active accounts in 190 markets and 24

> **PayPal could be a strong opponent to V/MC/AMEX in the next few years.**

currencies around the world, not to mention the broad reach of their parent company, eBay, which has over 16,000 employees and a market cap of over $32 Billion.

In Volume II of the SmartPhone Wallet Book series: *The Consumer Guide to the Services of Today and Tomorrow*, all of these companies and many more will be explored in detail to better understand their products, their value, and their future.

CHAPTER 6:

WHAT'S NEXT?

T he last Chapter took you around the world to illustrate various elements of the SmartPhone wallet in place and in use today. The Chapter prior to that explained the possible costs and cost savings that this new technology could offer.

However, there is a remarkably large gap between these real world examples, and the vision of the SmartPhone Wallet

This Chapter explores the elements that need to come together to make all of this work, namely:

1) **Industry Standards:** fragmentation and lack of international standards will slow adoption;
2) **Merchant Acceptance:** if the SmartPhone Wallet can't be used at enough retail locations, it will likely fail as a payment option; and
3) **Consumer Acceptance:** if every retailer on earth accepted a payment from a SmartPhone Wallet, but no consumers used it, then it will fail as a payment option.

Industry Standards

No worldwide standards exist today for any of the new format payment systems. If new formats to the current payment system are going to be widely available and widely adopted, or if the current format is to be made more efficient by the opportunity presented by the SmartPhone Wallet, then fragmentation needs to be avoided (or at least, not continued).

Both customers and merchants alike are more likely to adopt a unified industry.

Isis

A joint venture in the United States has provided a strong indication that mobile network operators are willing to work together. The joint venture announced at the end of 2010 by AT&T Mobility, T-Mobile USA and Verizon Wireless is called Isis™ and their primary charter is building a national mobile commerce network that aims to "fundamentally transform how people shop, pay and save," according to a press release[viii] from the company.

To do that, they will need to create an environment for V/MC/AMEX that will force them to change their business model or see their business erode.

The obvious route for Isis to take would involve an all-out attack on core the business model of MasterCard, Visa, and the other major players in the payments industry. That would certainly be an expensive battle. Entertaining -- but inefficient. Hundreds of millions of dollars would need to be spent trying to get the Isis brand to be as recognized and ubiquitous as Visa. Plus, Isis would also need to spend a considerable sum building the network infrastructure that would replace all of the functions that Visa handles today.

Do not be surprised if Isis actually supports Visa and MasterCard to speed adoption. On an even playing field over time, other payment solutions could manage money without the high rates charged to retailers by the payment system today. Which route will Isis take?

As the network that enables the SmartPhone to be trusted by other networks, Isis becomes the facilitator of other networks not the processor. In other words, Isis would be more than happy to complete a Visa transaction through its system if that is what the customer wants to do, and in all probability, these will be the majority of the transactions in the early growth stages of the company. But as Isis adds other payment solutions outside of the Visa or MasterCard networks, consumers may find themselves with attractive choices, and merchants may find themselves with a less expensive way to get paid.

This is important because if a retailer can save money (or, actually, keep more of the money that they are due) depending on which payment system you choose, the retailer is probably going to give the customer some incentives direct to their SmartPhone the moment before the consumer makes their choice.[ⁱ]

So, while Visa and MasterCard enjoy a position today with only a few competitors in the traditional payment schemes, they could find themselves with not only new competitors, but more efficient methods to transfer money from a consumer to a retailer.

Over time, as retailers encourage consumers to use other payment methods, and consumers actually do that, Visa and MasterCard face the difficult choice between changing their business model, or facing extinction.

GSMA / EPC

The GSMA and the European Payment Council[ix] have been advocating for the function of the Trusted Service Manager, for the last several years. We discussed the TSM as a security solution briefly in Chapter 4. Here is the background.

Although the original idea of the TSM was the brainchild of the GSMA, eventually they gained the support of the European Payments Council [x]– the organization responsible for making the payment system in Europe as efficient as possible. Both organizations agreed to work together on refining the roles and requirements of the TSM to facilitate a smart SIM card. This activity is one

> The GSM Association (GSMA) represents the interests of over 800 mobile network operators around the world.

example of the industry – in this case the mobile network operators and the European regulators – trying to get together and set a standard to avoid fragmentation.

If we can agree that standardization is critical than Isis represents the first significant attempt by mobile network operators to work together and find a standard that is optimally effective in the marketplace. If the Isis experiment works, it's powerful international partners, namely Vodafone and Deutsche Telekom, could potentially expand the standard of collaboration established in the United States by Isis.

Or, if you live in Canada, a company called EnStream, and its product "ZoomPass," is a collaboration between the mobile network operators of Canada, namely Rogers, Bell Canada, Telus, and Virgin mobile.

Perhaps Isis and EnStream will eventually deploy a common standard with common operational elements in the US and Canada, which would serve to be an example for the rest of the world.

Perhaps Safaricom, and other mobile network operators in Africa will come together and create a standard that the rest the world can use.

We know that standardization is possible, if everyone involved is motivated to get it done.

Merchant Acceptance

For mobile payments to reach their full potential, or any potential, the service needs to be available at typical brick and mortar stores.

The most widespread misconception about the potential of the SmartPhone Wallet is that it will primarily be a way to shop online. But, online retail purchases account for only about 4% of all retail transactions today, and will probably grow to about 8% by 2014. This growth rate is impressive, but the fact remains that brick and mortar retailers hold the key to the widespread adoption of the SmartPhone Wallet.

Even if every bank and credit union in the world enabled their customers to use a SmartPhone as a payment device, these applications would be virtually useless if you could not walk into your favorite store and buy something. Acceptance at brick and mortar retailers is key.

Transformative

What do brick and mortar retailers think about this technology and how do they think it will affect them in the future?

In a word, "transformative."

Retailers believe that soon a shopper will be able to walk down the aisles of their stores with their SmartPhone in hand and receive targeted messages, information about promotions, and even real-time coupons just by walking by a display on the floor. The ability for customers to use their SmartPhones to scan a barcode while standing in a store to check the price, compare it to other products, even search for current coupons is a reality today and possibly the norm in the very near future.

This is game changing.

There's a tendency to equate what's happening in mobile payments to the changes brought on by the Internet, but they're actually very different. The Internet dragged the customer along for a while — it was Evolutionary.

Here, consumers are leading the charge. This is Revolutionary.

"This is happening right before our eyes. We're standing on the brink of revolutionary change." Said Bernie Brennan, consultant and investor in retail-related technology and the co-author of *Branded! How Retailers Engage Consumers with Social Media and Mobility.*

Thanks to features like texting, GPS and Web browsing, mobile phones have quickly become a necessity many consumers cannot live without. Nearly three-quarters of retailers are exploring mobile strategies, according to a recent Shop.org survey. However, 62% of retailers have either not yet begun or are only in the early stages of planning their mobile strategy. According to the National Retailer Federation' Mobile Blueprint:

Mobile phones are changing the way retailers conduct business. Because they are always with us—and always on— they connect retailers to current and potential customers regardless of location or time of day. In 2015, shoppers around the world are expected to use their mobile phones to purchase goods and services worth close to $120 billion. In addition, payment for goods or services and money transfers initiated from a mobile phone will reach almost $630 billion by 2014, up from $170 billion this year. There is no doubt that mobile technology for retail is a hot topic worldwide.

New applications that enable mobile marketing and M-commerce are being published almost daily. There are applications that identify and reward shoppers as they walk into a store and others that allow consumers to give and get gift cards, receive personalized offers and monitor their loyalty programs from their mobile devices. There are tools that allow SmartPhone users to leverage social features as they shop, and others that allow consumers to make purchases using only their SmartPhone. And of course, there are applications that allow consumers to pay for their transaction.

Bottom line, the fact that retailers believe in the future of M-commerce, mobile payments and the like is good news for anyone who wants to use the services on a regular basis in the future. If the brick-and-mortar retail establishments were not actively looking at the sectors, the sectors themselves would have little chance of succeeding long-term.

PayPal and VeriFone

On October 27, 2010 PayPal - the largest alternative financial institution in the world - signed an agreement with VeriFone - the largest producer of credit card acceptance devices for retailers in the US.

If you read the press release[xi] announcing the partnership, it seems like an interesting niche service that is targeted at low-volume, small-business merchants.

VeriFone's PAYware is a piece of hardware that connects to an iPhone. The hardware is supported by a simple iPhone application, and the combination of application and hardware transforms an iPhone into a credit card reader. With the new PayPal partnership, PAYware will accept PayPal payments in addition to traditional credit-card transactions. The physical device provided by VeriFone will allow the user to swipe a card directly on their cell phone, and the cell phone will complete the transaction using the PayPal service. It may appear that this alliance is targeted at another card-to-mobile interface: Square, but that is not the main focus.

VeriFone may have something else planned.

A Closer Look at the Partnership

The partnership brings eBay-owned PayPal's merchant services to VeriFone's formidable product line. PayPal is the largest alternative payments company on earth with more than 87 million active accounts in 190 markets and 24 currencies around the world. PayPal is made up of three payment services: the PayPal global payment service, the Payflow Pro Gateway and Bill Me Later. Currently, all of these services are available only for online purchases, but try to imagine all of those forms of payment available at every retail store.

"Rapid adoption of mobile payments makes it possible to further align PayPal payments with traditional card-based transactions at the physical point of sale," said Jeff Dumbrell, VeriFone executive vice president.

Although Mr. Dumbrell politely used the word "...align...", what he is describing is really "competition." And the competitive advantage that VeriFone offers to PayPal is formidable; just look at their market penetration figures.

In 2009, approximately 13.1 Million POS terminals were shipped worldwide according to leading payment industry publication the Nilson Report. Of those, VeriFone shipped more than 3.1 Million terminals, or about 23% of the worldwide market. That makes VeriFone's market share of new shipments second in the world only to Ingenico (a French Company). In the US, VeriFone's share is even greater, with an overwhelming 48% share of new shipments, which translates to over one million units shipped annually, more than doubling the #2 player Ingenico. With a million new products shipped in the US annually, VeriFone could support a new technology and (assuming it could be priced competitively), deploy it across the US quickly. With their new VX product line, they are ready to do exactly that.

VeriFone is now supporting modules to its new VX Evolution line that do not inherently support NFC technology, but are designed to accept a module for various types of contactless payments. In an email interview with Pete Bartolik, VeriFone Media Relations, he said:

"Many of our products can be configured with contactless modules or accept a contactless peripheral. Earlier this year we announced a next generation product line, VX Evolution, with integrated contactless capabilities built in."

But how will a customer initiate a transaction at point-of-sale? How will a SmartPhone tell a VeriFone VX module to initiate a transaction through PayPal?

Bump

The quiet third party in the VeriFone/PayPal announcement is Bump Technologies. Bump Technologies wrote the software for the original application that swaps contact information by "bumping" two SmartPhones together.

VeriFone's PAYware is now enabled with Bump - the most popular way of establishing communication between two devices. So, how long before we see a new module for VeriFone's VX Evolution series that supports the Bump Technologies" product?

Probably in 2011 or 2012 at the latest.

If (when) that happens, merchants using the VX series would only need to purchase the "Bump Module" from VeriFone, and their system is upgraded.

Whiplash

For a merchant in the United States to support any of these new formats, they will likely be required to purchase new

POS equipment, which is a significant expense for retailers. However, if the current and emerging industries can demonstrate real standards that are likely to remain in place for all (or at least more than half) of the next decade, then merchants may be more likely to make the investment in these new technologies.

Merchants in the US are particularly concerned about an upgrade to their POS equipment because they only recently completed an upgrade to "contactless payments" which is likely to become immediately obsolete with the next equipment upgrade.

Here's why.

EMV, NFC and Bump

EMV is a credit card security standard created in Europe. It has less than one-fifth of the fraud occurrences when compared to the US 'magnetic strip' system. Although the US just upgraded to contactless, it is now more secure than magnetic strip with one extra layer of security, but it is still not as comprehensive as EMV. And, an EMV upgrade will definitely include hardware.

Since NFC is a small radio signal, and a 'Bump' initiated transaction requires a piece of equipment that can also 'sense' a bump, hardware upgrades are required at the retail locations to enable these technologies.

This will cost a lot of money so before they make the choice to upgrade they will want to make sure that fragmentation has ceased and standardization is in place.

Customer Acceptance

This is where you come in.

Given the flood of activity and offers in the marketplace, you are more likely than not to participate in some SmartPhone Wallet offering. As this book has explored, a format war causes people to wait for a winner before choosing to participate. Unfortunately, that is just like not voting in a political election and hoping the people that <u>do</u> vote make the right choice.

You must participate to ensure that the best product for you has the support of the market. The 'market' starts with you.

CONSUMER TIP

Participation will be better than a 'wait and see' approach to the SmartPhone Wallet products of tomorrow

The Point of Sale

As we discussed earlier, the industry emerging to support mobile payments (among other things) is called Near Field Communication, NFC, and it will enable the mobile payments industry to be user friendly.

A recent analysis from Juniper Research regarding the NFC opportunity forecasts that:

"...the application of NFC as a mobile retail marketing tool via coupons and smart posters will support the growth of NFC mobile payment transaction values from $8B in 2009 to $30B within three years."

NFC Stickers

NFC stickers are, well, exactly what they sound like: a sticker. On the surface, they are not substantially different from each of the 300 princess images in a sticker book that I just bought for my three-year-old. Of course, an NFC sticker is more than that because it contains a radio–frequency (RF) chip inside the sticker itself. The sticker is also a lot thicker, generally square in shape, and has a different texture than Cinderella, but the RF chip is really the significant difference.

There are many examples of NFC Stickers being used in the marketplace today, including Bling Nation, First Data's Go-Tag, Zoompass, Discover's Zip and a variety of others.

Most of these products use a small microprocessor that is embedded inside the sticker that is very similar to the microprocessors that are embedded in most credit cards and debit cards issued after 2007.

The real benefit of the NFC sticker is that it is inexpensive, and of course it will work with any SmartPhone. Of course, the NFC sticker is not integrated into the functionality of your SmartPhone in any way, and therefore does not take advantage of any of the *smarts* of your SmartPhone.

You could literally put the NFC sticker on a block of wood if you somehow find that more convenient, and it would function exactly the same as if you had put it on the back of your SmartPhone.

Such is the excitement about the potential of NFC. Vendors are developing and launching a variety of interim solutions such as stickers to get NFC to market faster on existing phones rather than waiting for new NFC enabled phones.

NFC report [xii]author Howard Wilcox stated: "Many people focus on the use of NFC for payments but in fact it is poised to revolutionize the way many people shop too. The ability to tap smart posters and receive coupons and product information also presents new channels to market for merchants. Whilst vendors see widespread availability of NFC phones in future, the jury is out as whether interim solutions will attract users or actually have a detrimental effect." Further findings from the NFC research include:

• NFC devices were shipped commercially in 2009 and the market will ramp up from 2011.

• NFC/FeliCa payments are already established in Japan but by 2014 North America and Western Europe will be experiencing high growth.

• By 2012, NFC global gross transaction value will exceed $30B.

Embedded NFC

The third way of enabling NFC communication in a SmartPhone is to have the manufacturer install both the NFC chip and the corresponding antenna into the phone itself. In this case, of course, the NFC technology is fully integrated into the SmartPhone. An example of this is the C7 SmartPhone from Nokia. A phone with integrated NFC capabilities allows the SmartPhone's operating system to interact with NFC communication chip. This is substantially better than both the sticker, and the add-on device because it is operated by the phone's operating system, and any application that is on that SmartPhone can use the NFC functionality to communicate with another enabled device.

Another upside of this solution is the relatively low incremental cost compared to the add-on device. As of the end of 2010, there are relatively few NFC enabled phones on the market, and to upgrade to this type of solution would require that the consumer buy a new SmartPhone.

UICC / SIM Card Enhancements

As described in Chapter 1, several types of add–on devices can upgrade a normal SmartPhone, or even some other types of mobile phones, to be an NFC phone.

UICC (SIM) cards can also do other things, such as carry an NFC chip with a small antenna that the NFC sticker uses. However the advantage in putting the chip onto a SIM card allows the NFC functions to integrate with the phone. Although this solution costs substantially more than the NFC sticker, add-on devices provide the same integration into the software of the SmartPhone which allows consumers to manage the NFC functions through an application on the phone, which is one of the formats of the SmartPhone Wallet.

According to a company called Device Fidelity, approximately 65% of phones in use worldwide can accept an add-on as either a SIM card or a similar removable card called a Micro–SD card.

Bump

Although we already discussed the technology above in the context of merchant acceptance, it also matters that consumers embrace the technology, so here is some customer-focused detail.

Bump involves a transaction that begins between two enabled devices by literally tapping a SmartPhone to a compatible device -- like another compatible SmartPhone, or a VX terminal -- that uses the phone's sensors to literally "feel" the bump. Then, it sends that information into the Internet to look for a matching bump from phones around the world and pairs up phones that felt the same bump. Then, Bump can route information between the two phones that match.

As of the end of 2010, Bump supports iPhone, iPod touch, iPad, and Android. They are working to bring Bump to other platforms soon. While several other phones that include the sensors that Bump needs to feel bumps exist currently, in the next few years the vast majority of new phones will have these sensors.

Bump plans to have an application for all SmartPhones as they become available.

Opting Out

It could be two decades before the last magnetic strip reader is unplugged, maybe more. If you want to ignore new technology and keep using your magnetic strip credit card to buy things, you probably can do that for at least that long.

So, deciding not to decide is -- as usual -- an option.

CHAPTER 7:

CONCLUSION

Given the opportunity, most of us would choose to
end hunger and achieve world peace above all other human
endeavors. But of course, history has shown this to be a
nearly impossible goal.

Nearly impossible.

No; neither the SmartPhone nor the SmartPhone Wallet is
the sole answer to world peace. However, the SmartPhone
Wallet has a unique and remarkable opportunity to actually
play a part in the peace that could be achievable in this
century. This "big picture" connection has to do with hunger
and distribution of wealth.

To put it simply, this technology can move buyers and sellers closer together. When that happens, the possibility exists for fewer "middlemen" who profit from simply moving money around.

Moving money costs money. As you have seen in this book, for example, when you move money from your credit card account to a merchant's bank account, the system that does that takes around 3% of the transaction amount.

Consider for a moment the possibilities discussed in this book -- namely how the SmartPhone Wallet can provide remarkable control over how you use your money, which you send your money too and how you receive it.

If a provider of an intellectual product can be paid directly by the user of the product -- or at least with the fewest possible number of "middleman" who profit from simply moving the money around, then more money goes to the actual provider.

Here's an example:

In the Spring of 2006, Todd Grossman, a New York City resident, was invited to join an expedition to the summit of Mount Kilimanjaro in Africa. Todd booked his expedition through with a New Jersey-based company and proceeded to Tanzania for the trip of a lifetime. What he didn't realize until after the trip was underway, was that the tour operator was merely a broker; a middleman who got paid for collecting and redistributing the money.

As Todd ventured up Mount Kilimanjaro, he became fast friends with the lead guide, Jonas Loiruk Eliau, of Ilkrevi Village just outside of Arusha in Tanzania. Jonas was in his twenties and managed a team of 20 porters who carried supplies up the mountain for the participants. As they climbed, Jonas and Todd discussed a wide range of topics, including the value Tanzanian people place on education, the economic plight of his country, and how the problems associated with rampant poverty and corruption. Eventually, Jonas mentioned how much money (or, rather, how little) he was earning for organizing and leading the tour.

Todd calculated how much he had paid, multiplied by the number of people who were also paying members of the climb, and estimated the revenue the tour generated for Jersey-based purveyor. Astonishingly, he calculated that approximately 75% of the tour's revenue stayed in the US in the form of pure profit to the broker. Todd quickly realized that with a mobile phone and a mobile Internet access, Jonas needed little more than some advice and encouragement to market his services directly. Over the next year, Todd helped Jonas create a website and establish basics financial exchange services. Soon, Jonas and his colleagues were able to charge climbers less money, and earn, not just a little bit, but substantially, more money. Today, both Todd and Jonas help students learn the value of one to one economic connections through the Kilimanjaro Education Foundation.

> One of the goals of the Kilimanjaro Education Foundation is to teach students how to build one to one economic connections.
>
> Source: www.kef4kids.org

Here, you're probably saying, "Of course. Using the Mobile Phone to access the Internet to sell directly to people, and cut out the middleman, is a much more efficient way to keep more of the money that workers earning." But it goes farther than that. Because Jonas earns more money, and the twenty porters now under his employ also earn more money, they spend more money. And that benefits the community in which they work. All

Todd and Jonas did was to realize that the disproportionate profits taken by the US company could be erased based on modern mobile and Internet technology.

The SmartPhone Wallet has the potential to take this concept a giant step forward, because in an environment where a worker can only be paid in cash, and does not have a bank account to deposit the funds into, that worker must rely on some other service to find a way to distribute their earnings in the form of cash. In developed nations, there are enforceable laws that prohibit extraordinary fees to be charged for simply cashing a check, or advancing money. However, in areas where the service is needed most -- the developing nations of the world -- few such laws exist, and those that do are difficult or impossible to enforce.

The SmartPhone Wallet could change much of that by giving the workers in this example the ability to get paid directly by his employer depositing funds into an account that can be managed by his or her

> **In 2012, over 70% of the people in the world who do not have a bank account use a mobile phone.**
>
> -World Bank's Microfinance Group

SmartPhone. The World Bank's microfinance group reports that the number of people who lack a bank account but have a mobile phone will reach 1.7 billion in 2012, about 70 percent of the entire unbanked population worldwide[xiii].

So, it is essential that we -- the consumers -- find and use those new SmartPhone Wallet products and services that are optimally efficient, provide the most cost savings, and ultimately make it easier for buyers to pay sellers directly.

The success of the programs in Africa to date, are only a start. Unfortunately, each program in each country, provided by each carrier, are all fundamentally different, and have different features, benefits, costs, and processes. In other words, there are few, if any standards for SmartPhone Wallet programs in the marketplace today anywhere in the world.

This ad hoc system is not sustainable.

If the Smartphone Wallet is to reach its full potential around the world, and have the economic affect on developing nations that is clearly possible, then standards need to be developed and applied worldwide. That does not mean that the future SmartPhone Wallet product offered in Tanzania should have the same features and benefits of the SmartPhone Wallet products offered in Hong Kong and New York; rather a standard of operations, security, and integration must be adopted to avoid a crippling level of operational complexity that would happen if every program in the world simply worked differently.

You may have noticed this book has attempted to remain neutral regarding which format of the SmartPhone Wallet you should choose, or which specific products, services, and providers will give you the best advantages.

Instead, the philosophy of this book has been to advocate an open mind, a willingness to try these products as they are presented to you, and a willingness to switch from one product to another in an attempt to understand which is best.

This new industry will happen. Whether it works the same way as credit cards work today or whether the payment system changes entirely; the SmartPhone Wallet is becoming a reality.

Some of the products and services are here already, and they will only grow in number, volume, and exposure. You will hear about them, will be presented with offers to use them, and you will have to decide whether to participate or not.

Participate. Be flexible. Stay informed. Try different products. Be mindful of the privacy policies for everything you use.

And remember, if we can all get this right, and let the best products in the marketplace survive and thrive, there exists the potential for changing the world simply by bringing buyers and sellers closer together.

About the Author

David W. Schropfer is an international business leader with almost two decades of management experience ranging from telecommunications to payment systems. He helps companies meet their international objectives by leading collaborative business teams, creating new business strategies, identifying and negotiating strategic partnerships. Since co-founding the Luciano Group in 2005, he has been responsible for strategic development and negotiation for the firm and its clients in the areas of mobile payments, partnership agreements, and interconnections.

Prior to founding The Luciano Group, David was Senior Vice President with IDT Telecom, and head of International Business. In this role, David created over six million new EBIT dollars annually by building wholesale voice/data interconnections, prepaid Visa debit card programs and prepaid calling card programs with mobile carriers, fixed line carriers, Second National Operators and other partners

Growing experienced teams and negotiating partnerships has helped David grow businesses in other positions, including his CEO role at SoftZoo.com, where he built his C-level team, recruited a top-flight Board, and raised millions of dollars of investment capital. Earlier, David was the Business Development Officer for the MVNO division of Capital One. In this role, David was responsible for applying Capital One' credit card business model to the mobile resale business. He established partner relationships, designed and negotiated partnership agreements, developed complex business models and effectively lowed the cost structure.

Since graduating Boston College, David earned an Executive MBA from the University of Miami. He lives in New Jersey with his wife and children.

End Notes

[i] Sources include Media College (www.mediacollege.com), CNet (www.cnet.com) and others.

[ii] Apple and the Apple logo are registered trademarks of Apple Inc. Microsoft and the Microsoft and Windows logos are registered trademarks of Microsoft Inc.

[iii] As of December, 2010 when this book was published, a significant number of regulatory efforts were underway in various parts of the world which may dramatically change the rules and limits associated with both Interchange and Transaction fees for credit and debit cards.

[iv] The Luciano Group Press Release- *Isis Market Opportunity Will Exceed $50 Billion by 2014*– December 2010

[v] PC World Magazine, Aug 11, 2010 , *Smartphone Security Thwarted by Fingerprint Smudges*, by Tony Bradley

[vi] GSMA Trusted Service Manager Recommendations, January, 2010.

[vii] Although the FeliCa system was proposed to the International Organization for Standardization (ISO) for the international NFC standard (known technically as ISO/IEC 14443), it was rejected. So, the terms "NFC" and "FeliCa" are not the same thing.

[viii] Isis Press Release, *AT&T, T-Mobile and Verizon Wireless Announce Joint Venture to Build National Mobile Commerce Network,* November 16, 2010

[ix] EPC – GSMA, Trusted Service Manager Service Management Requirements and Specifications. Doc: EPC 220-08, Version 1.0. January 2010

[x] According to its web site, "The European Payments Council (EPC) is the decision-making and coordination body of the European banking industry in relation to payments." http://www.europeanpaymentscouncil.eu/

[xi] VeriFone Press Release: *VeriFone Extends Mobile Payments Acceptance with PayPal*, Oct. 26, 2010

[xii] Press Release: NFC Mobile Payments to Exceed $30B by 2012, September 2, 2009:

[xiii] Wireless Week Magazine, *KDDI Invests $22M in Mobile Payments,* By Maisie Ramsay, December 13, 2010

13572183R00082

Made in the USA
Charleston, SC
18 July 2012